Classroom Companion: Business

The Classroom Companion series in Business features foundational and introductory books aimed at students to learn the core concepts, fundamental methods, theories and tools of the subject. The books offer a firm foundation for students preparing to move towards advanced learning. Each book follows a clear didactic structure and presents easy adoption opportunities for lecturers.

More information about this series at http://www.springer.com/series/16374

James E. Sallis • Geir Gripsrud •
Ulf Henning Olsson • Ragnhild Silkoset

Research Methods and Data Analysis for Business Decisions

A Primer Using SPSS

Springer

James E. Sallis
Department of Business Studies
Uppsala University
Uppsala, Sweden

Geir Gripsrud
Department of Marketing
BI Norwegian Business School
Oslo, Norway

Ulf Henning Olsson
Department of Economics
BI Norwegian Business School
Oslo, Norway

Ragnhild Silkoset
Department of Marketing
BI Norwegian Business School
Oslo, Norway

ISSN 2662-2866 ISSN 2662-2874 (electronic)
Classroom Companion: Business
ISBN 978-3-030-84420-2 ISBN 978-3-030-84421-9 (eBook)
https://doi.org/10.1007/978-3-030-84421-9

Translation from the Norwegian language edition: Metode og dataanalyse: Beslutningsstøtte for bedrifter ved bruk av JMP, Excel og SPSS, 3. utgave by Ulf Henning Olsson, et al., © CAPPELEN DAMM AS 2016. Published by Cappelen Damm Akademisk. All Rights Reserved.

This Springer imprint is published by the registered company Springer Nature Switzerland AG.
The registered company address is: Gewerbestrasse 11, 6330 Cham, Switzerland

Preface

We live in a world characterized by big data, artificial intelligence, robotics, and automated analyses. This is quickly changing the role of managers, and if they are not careful, making them redundant. Decisions based on gut feelings are giving way to data-driven decision-making. Today's manager must understand the power of data analysis and be able to discern the quality of the analysis. Today's challenge is not about the amount of data; it is about the quality and processing of the data. Data analyzed with the wrong techniques will generate bad recommendations, no matter how small or big the database. Data alone does not create value. Worthwhile insights are the result of skilled analysis and interpretation.

We, as individuals, create more data than ever before. Digital footprints predict future behavior, even without the actual identity of individuals. We see social media creating immense opportunities for qualitative and quantitative studies.

This book is written for budding decision-makers in business and organizations: for people who understand the power of data and want to learn the basic analysis techniques to explore, test, and validate options. The book describes the research process from both qualitative and quantitative perspectives. Using hands-on examples with QDA Miner Lite for qualitative analysis and IBM SPSS Statistics for quantitative analysis, the book teaches the basic skills to enter the world of data analysis.

Geir Gripsrud, Ulf H. Olsson, and Ragnhild Silkoset are Professors at BI Norwegian Business School and James E. Sallis is Professor of Business Studies at Uppsala University. All four have extensive experience in applying and teaching qualitative and quantitative research methods to a global audience.

Web support including datasets, short videos demonstrating SPSS, and other literature can be found on Professor Sallis' website at Uppsala University: https://uppsala.instructure.com/courses/39264.

Other product and service names might be trademarks of IBM or other companies. A current list of IBM trademarks is available on the Web at "IBM Copyright and trademark information" at www.ibm.com/legal/copytrade.shtml.

Uppsala, Sweden
Oslo, Norway

James E. Sallis
Geir Gripsrud
Ulf Henning Olsson
Ragnhild Silkoset

Contents

Part I

Designing the Study

Research Methods and Philosophy of Science

1

Contents

1.1 Introduction

Big data, SoLoMo (social, local, and mobile), Screen Scraping, ClickStream, Passive Data Collection, Algorithms, Network Intelligence, Real Time, WebCrawling, Data Mining, Dashboards, Blog Mining, Neuromarketing—never before have data analytics been subjected to such intense creativity and innovation. Never before has it been more important to return to the roots, to the scientific foundation of how to interpret, understand, and rely on data analysis. Making flawed decisions based on huge amounts of data through automated data processing can be more damaging than not making decisions at all.

In this book, you will learn the scientific basis for doing research and conducting data analysis for business. This includes procedures, interpretation, and application. A "method" means a planned approach. Deciding on a method requires first understanding goals: what is to be achieved. Given the business context, the book is written to have relevance for organizational decisions. There are countless decisions, each with characteristics pertinent to method choice. Accessible resources, both in terms of time and money, will also influence method choices. However, are there other things that matter?

J. E. Sallis et al., *Research Methods and Data Analysis for Business Decisions*, Classroom Companion: Business, https://doi.org/10.1007/978-3-030-84421-9_1

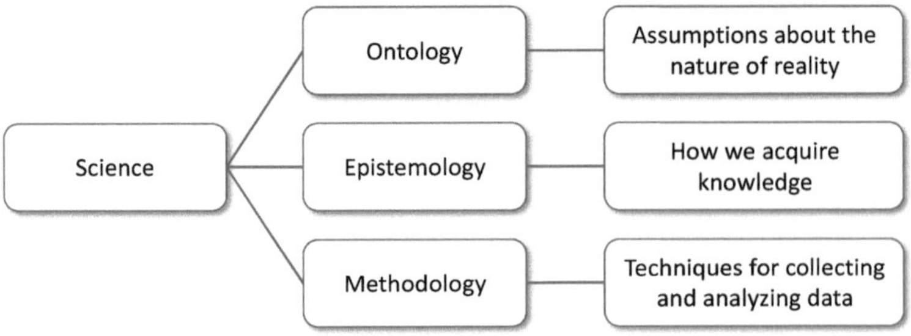

Fig. 1.1 Three assumptions in science

At first glance, it may seem rather unproblematic to acquire better knowledge. Still, there are many hidden dangers. When pondering what it means to acquire better knowledge, we encounter philosophical choices. Explicitly or implicitly, we take a position regarding two concepts within the philosophy of science: ontology, which addresses philosophical assumptions about the nature of being (to be, or not to be), and epistemology, which is how we acquire knowledge of reality. It is the way by which we know things. On a more concrete level, "the way we know things" refers to research methodology. It is a combination of methods and techniques to investigate something (see Fig. 1.1).

Ontology, epistemology, and methodology are connected. The assumptions we make regarding how we perceive reality provide guidance on how we acquire knowledge about reality, and thus, what kind of methods we apply.

1.2 Two Alternatives: Constructivism Vs. Positivism

We can gain knowledge about a company by reading annual reports, analyzing accounting figures, studying the market for its most important products, and so on. In this form of learning, we keep a distance from the company, or companies being studied. We assume that they exist *independently* of us, are well defined, and are not affected by the fact that we investigate them. This is the ontological basis for the *positivist paradigm.* Positivists perceive knowledge as virtually *objective.* Positivism, as a philosophic scientific paradigm, originates from the natural sciences. Arguably, it cannot simply be transferred to the social sciences since what we study, our phenomena, are fundamentally different. A diametric alternative to positivism is the *constructivist paradigm.* Constructivists perceive knowledge as *subjective*. Phenomena, like companies, are not clearly defined and are *not independent* of being observed. We gain knowledge through interaction and reflection.

Methodology stems from epistemology, which stems from ontology. Both constructivism and positivism have practical relevance for studying companies. Yet, we face a dilemma by embracing two diametrically opposed ontologies. This book takes a *pragmatic approach* combining both paradigms, though we admit to leaning toward positivism, manifested in the emphasis on quantitative data analysis.

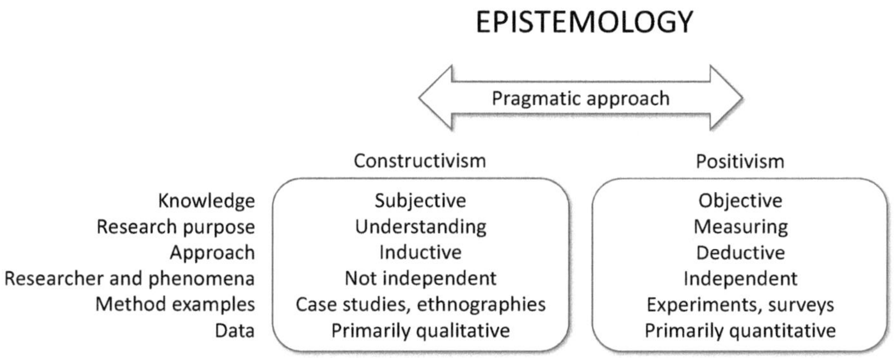

Fig. 1.2 Constructivism and positivism, and the pragmatic approach

In a business context, where we want to acquire knowledge that is relevant to a company, under certain circumstances it may be necessary to enter the company and participate in their business activities. We talk to workers, enquire about their attitudes and beliefs in various contexts, interpret their norms, and experience how tasks are performed. This sort of research provides deeper insights into how the business works than what we would obtain through typical positivist approaches. With a positivist approach, we clearly define concepts, typically stay outside of the company, and collect data from sources like employee questionnaires or publicly available financial reports.

Social anthropologists who study foreign cultures primarily use a constructivist approach to gain knowledge. Students doing internships in companies, though not explicitly doing research, apply similar forms of knowledge acquisition to interpret and understand what is happening in the company.

Although constructivism and positivism are based on different assumptions and have different strengths and weaknesses, they complement each other. An analogy is the increasing emphasis on student internships connected to academic studies. The knowledge acquired from learning by doing during the internship supplements learning from books and lectures. Together, they provide a whole that neither knowledge source could provide alone. The issue being faced determines the choice between paradigms. A constructivist approach is well-suited to *understanding*, for example, how well a corporate culture is functioning, whereas a positivistic approach is well-suited to *measuring*, for example, the relationship between the price of a product and its level of sales. If we are keen to investigate the occurrence and meaning of references to a brand in social media, both approaches may be relevant. However, the research questions we pose within the two approaches will be different, as will the results.

The two paradigms differ in ontology and epistemology, which has implications for the research methodology and analytic techniques (see Fig. 1.2). In the positivist paradigm, where *quantitative* techniques are common, emphasis is placed on *measuring* relationships between variables. In the constructivist paradigm, the emphasis is on *understanding* social phenomena through *qualitative* methods. The priority is to understand phenomena as a whole in relation to other phenomena, so deconstructing phenomena into smaller parts that are measured through, for example, structured questionnaires is not appropriate. Extensive in-depth interviews, as an example, are better suited for building understanding.

William B. Cameron eloquently expressed the tension between the paradigms:

> Not everything that can be counted counts,
> and not everything that counts can be counted (1963, p. 13)

As a sociologist, he was addressing the fact that not all data can be quantified. In the positivist paradigm, with its emphasis on quantitative methods, there is a tendency to disregard what cannot be counted. However, not everything that counts can be counted! In the constructivist paradigm, there is a suspicion that positivists lie with statistics. We propose a truce. With a pragmatic approach, the premise is neither only about hypothesizing the connection between variables (positivism) nor only interpreting social phenomena (constructivism). The pragmatic approach starts with the need to understand and measure phenomena. Very simplistically, qualitative research unveils the existence of something, while quantitative research counts how much.

Three typical role models, physicist, psychiatrist, and physician, illustrate the two paradigms and the pragmatic approach. The physicist is a disciple of the positivist paradigm, while the psychiatrist is a disciple of the constructivist paradigm. Meanwhile, the physician uses a pragmatic approach applying mixed methods to form insights from several scientific domains to explore and understand phenomena. The physician is primarily concerned with making the correct diagnosis and treatment decisions. Combining insights from both quantitative sources (e.g., physical tests like blood pressure) and qualitative sources (e.g., social interaction with the patient) is useful, if not paramount, in making the correct decisions.

In Fig. 1.2, we provide some characteristic features of constructivism and positivism, and show how the pragmatic approach spans both paradigms. We show that the purpose is different in each approach. In a sense, we can say that the starting point is a "problem" in all three types of research. However, the nature of the problem is different. In the positivist paradigm, usually one or more hypotheses are derived from theory, and the question becomes how to test the hypotheses in the most efficient and robust way. In the constructivist paradigm, we want to form an understanding for a process or interpret the meaning or content in a process. The question becomes how to achieve the best possible understanding when we consider that we ourselves participate in the process. In the pragmatic approach, we have to make decisions, and the problem is which decisions to make. This is a typical scenario for managers. They have to "make a diagnosis" and "prescribe the medicine" to the organization. Though this book emphasizes the positivistic paradigm, we recognize, respect, and extoll the pragmatic mixed-methods approach to business research and decision-making.

1.3 Falsificationism, Theories, and Hypotheses

The word "fact" in social sciences comes with challenges. On the spectrum between constructivism and positivism, how can researchers be so certain as to say that something is a fact, or that something is proven? The short answer is, they cannot.

Rooted in the work of Karl Poppèr (1902–1994), *falsificationism* purports that a theory is only scientific when it is empirically testable, or in his words, it can be falsified. From this perspective, a theory is sustained until an observation demonstrates it to be false. While one observation may be a coincidence, several observations build a chain of evidence either to support or refute the theory. This same logic holds when considering hypotheses. For any hypothesis, there must be an alternative hypothesis. In inferential statistics, the null hypothesis is, by default, that there is no relationship between variables. The alternative hypothesis is, then, that there is a relationship.

Social scientists adhering to the falsificationist doctrine will posit (hypothesize) an empirically testable hypothesis, with the objective of testing theory. Social science is in a constant cycle of inductively developing theory from observations, and deductively testing theory with observations (see Fig. 1.3). Popper argued that induction is not science in so far as it builds conclusions from observations. No matter how many times a phenomenon is observed, the conclusion from observation, the theory, is not science until it is expressed in such a way that it can be empirically falsified.

Falsificationism is the underlying perspective on science within this book. Induction is a valuable part of the scientific process, and as such, we view induction as a scientific method. Induction and deduction are simply two complementary processes. To argue that one is superior to the other is like saying *breathing in* is superior to *breathing out.* After all, breathing in feeds oxygen to the body. However, try doing it without breathing out!

A hybrid view of the scientific process is *abduction. Abductive reasoning* starts from observation and seeks the most likely explanation (theory). In line with falsificationism, it does not claim the explanation to be factual. The pragmatic

Fig. 1.3 Theory development

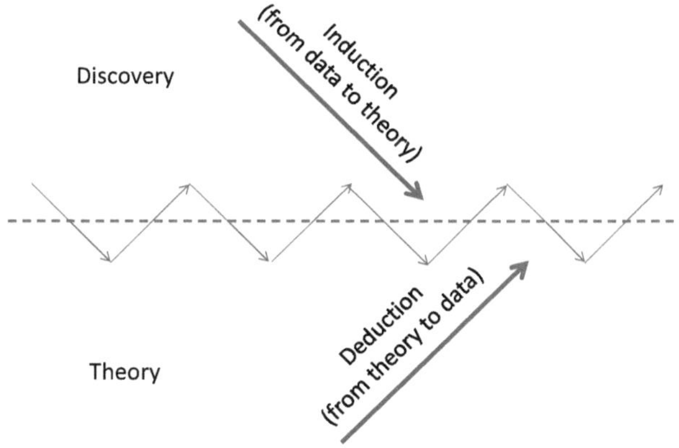

approach described in the previous section is a form of abduction. The physician makes the best possible diagnosis based on available observations. If confronted with additional information that leads to other conclusions, the physician will revise the diagnosis, which in our case would be the theory.

1.4 Decision Processes

Managers constantly make decisions in different situations. Some decisions are routine, while others have strategic implications and are likely infrequent. Routine decisions typically follow fixed procedures; something happens that triggers a decision. For example, an inventory system tracks the rate of sales relative to inventory for a specific item. When the inventory reaches a specified minimum level, an order decision is triggered. We often see inventory systems when we shop online. We click on an item and get a count of how many are in stock and buying options (see Fig. 1.4). Our purchase choice will trigger inventory decisions with the retailer.

When managers in a company decide whether to enter a long-term agreement with, for example, a specific supplier, the decision is not routine. In such a situation, managers need to gather information to form a basis for the best possible decision. These are the sorts of decisions we concentrate on in this book. These decisions are

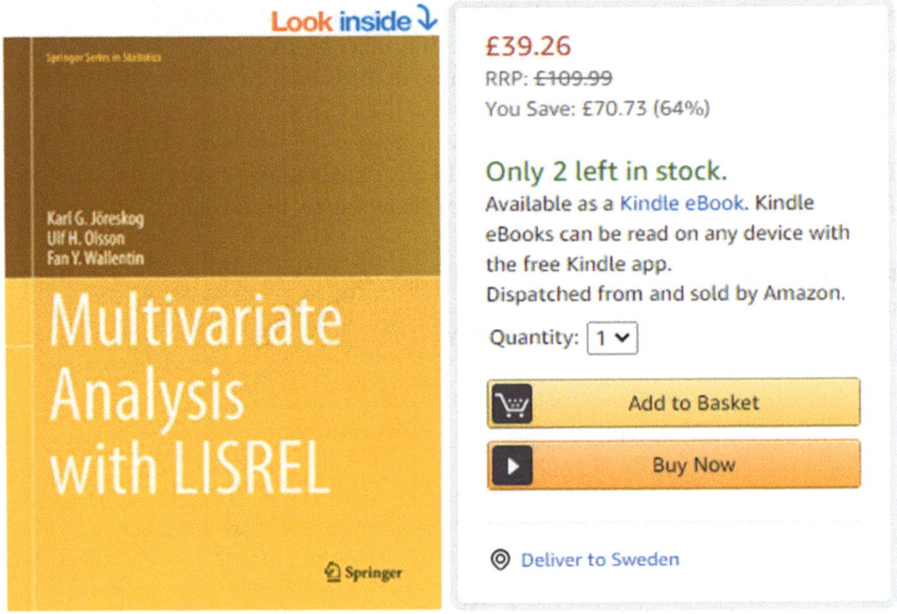

Fig. 1.4 Routine decisions

often made in a highly uncertain environment where inputs and results are ambiguous.

Examples of these sorts of strategic decisions are found across all functional areas of a company:

- *Purchasing*: Do we enter purchasing agreements with other companies?
- *Finance*: Do we raise money through stock issues or bank loans?
- *Personnel*: Should we implement or expand an employee bonus system?
- *Marketing*: Should we shift budgets toward social media?
- *Accounting*: Should we externalize our accounting functions to an accounting firm?
- *Research and development*: How should we organize product development?
- *Strategy*: Which international markets should we enter with new products?

One can ask how such questions arise and what puts them on the agenda for individual companies. There may also be reason to ask how the actual decision-making process takes place.

There has been a lot of research on decision-making, both within psychology and within management and strategy research. Part of this literature is primarily concerned with *normative models*, that is, how decisions should be made. There are also many *descriptive models* of how decisions are actually made. Eisenhardt and Zbaracki (1992) argue that there are three fundamental paradigms that apply to the research on strategic decision-making. These are:

- The decision-making process characterized by rationality and bounded rationality,
- The decision-making process characterized by politics and power,
- The decision-making process characterized by ill-defined preferences and fluid participation.

The first approach focuses on how an individual weighs alternatives to arrive at a choice, whereas the other two are concerned with how individuals and groups interact to arrive at an outcome. The first approach divides into rational and "irrational" models. We call this first main approach individual models (see Fig. 1.5).

1.4.1 Models of Decision-Making

In the *classic rational model*, *economic man* as the decision-maker has a clear goal, perfect information about all possible options, can assess all possible consequences of these options, can rank all options based on the consequences, and can choose the option that ranks highest. When the choice can result in different outcomes, and each outcome has a given probability, the decision-maker will choose based on a combination of probability and the utility (total satisfaction) of each possible outcome. This is called *expected utility theory* (Von Neumann & Morgenstern, 1944). In

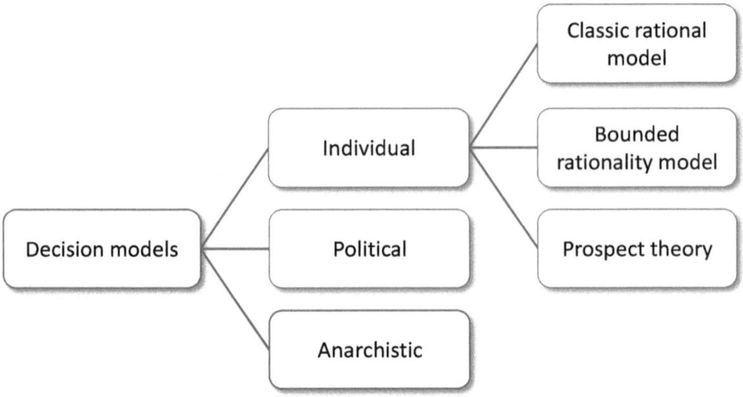

Fig. 1.5 Decision models

simple terms, a decision-maker will always choose the option that gives the highest expected benefit. Two distinctive assumptions of this model are that (a) the decision-maker only considers the statistical probability of the various outcomes, which implies risk neutrality, and (b) the decision-maker is not influenced by the way the options are presented. While the original theory was formulated using objective probability (based on data), later versions included subjective probability (intuition or judgment-based). It is a normative theory intended to suggest how decisions should be made, rather than a descriptive theory of how decisions are actually made (Plous, 1993).

The *bounded rationality model*, on the other hand, attempts to describe how individuals actually decide. It does not assume that decision-makers have perfect information about all alternatives and their consequences. Herbert Simon (1955) was central to developing this model. He argued that decision-makers, as a rule, seek an alternative that is satisfying and suffices (is good enough), or what he called, "satisficing." When a decision-maker finds a satisfactory option, he or she chooses it without seeking better possible alternatives. This means that the order in which alternatives are discovered or presented affects the choice, which contradicts the assumptions of the classic rational model.

The third model we consider, developed by Kahneman and Tversky (1979) and called *prospect theory*, describes how individuals have different starting points and make choices under conditions of uncertainty. Decision-makers start with a set of options, called prospects. Each prospect has potential losses or gains. The decision is a function of weighing the gains against the losses from the given starting point, which is an important distinction from expected utility theory. People are more influenced by potential losses than gains, even when the expected value is identical. Experiments show that most people prefer a certain prize of $500 to a 50% chance of winning $1000, while most people prefer a 50% chance of losing $1000 to a certain loss of $500. Expected utility theory values these two outcomes as equal, whereas prospect theory recognizes human nature's aversion to losses. This implies that the way gains and losses are framed (presented) influences choice (Plous, 1993). Richard Thaler won the 2017 Nobel Prize for his theory of nudging (Thaler, 2015). Nudging could be described as subtle framing, a little push, on the decision process that can have substantial ramifications for choice.

1.4.2 Models of Politics and Power

The second main type of decision models focuses on organizational decision-making embedded in the context of politics and power. Decisions are reached through a struggle between different interest groups in the organization (Pfeffer & Salancik, 1978). This struggle is similar to the conflicts in national politics. Decisions are often the result of conflicts where different interest groups apply pressure, and the outcome depends on who has the most power. This perspective says little on how the individual interest groups choose the alternatives they prefer, though it is reasonable to assume that they are a result of individual decision processes taking place prior to the political processes.

1.4.3 Models of Ill-Defined Preferences and Fluid Participation

The third main type of decision models consists mostly of so-called garbage can models (Cohen, March, & Olsen, 1972). Decision-making processes are highly unstructured and depend, to a large extent, on coincidence. Different people participate in the process at different times, they have no clear goals, and they are often simply trying to reach any decision. This paradigm is a reaction to the rational decision-making models. It asserts that in reality, problems, goals, information, and outcomes are often not clear, thus identifying relevant solutions and who should participate in decisions is ambiguous.

1.5 Summary

Our point of departure was that non-routine decisions in companies and organizations generate a need for information and knowledge. Knowledge can be generated in many ways. From a scientific perspective, we distinguish between philosophical assumptions about how reality is (ontology) and how one can acquire knowledge about it (epistemology). Positivism and constructivism are two extremities on a continuum of how researchers view reality and how they acquire knowledge. Positivism emphasizes measuring, while constructivism emphasizes interpreting and understanding. In this book, we promote a pragmatic, mixed-methods view—in line with a physician's role—where the task is to make a diagnosis and validate options for making decisions and taking actions. There is no easy explanation for how decisions are or should be made in companies and organizations. This book is written to support the individual decision-maker. We ignore politics, power, and garbage cans. We assume some degree of rationality in the individual who, when faced with a decision, needs reliable information.

References

Cameron, W. B. (1963). *Informal sociology: A casual introduction to sociological thinking*. Random House.

Cohen, M. D., March, J. G., & Olsen, J. P. (1972). A garbage can model of organizational choice. *Administrative Science Quarterly, 17*(1), 1–25.

Eisenhardt, K. M., & Zbaracki, M. J. (1992). Strategic decision making. *Strategic Management Journal, 13*(S2), 17–37.

Kahneman, D., & Tversky, A. (1979). Prospect theory: An analysis of decision under risk. *Econometrica, 47*(2), 263–291.

Pfeffer, J., & Salancik, G. R. (1978). *The external control of organizations: A resource dependence perspective*. Harper & Row.

Plous, S. (1993). *The psychology of judgment and decision making*. McGraw-Hill.

Simon, H. A. (1955). A behavioral model of rational choice. *Quarterly Journal of Economics, 69* (1), 99–118.

Thaler, R. H. (2015). *Misbehaving: The making of behavioural economics*. Allen Lane.

Von Neumann, J., & Morgenstern, O. (1944). *Theory of games and economic behavior*. Princeton Univ.

The Research Process and Problem Formulation

2

Contents

2.1 Introduction

What insights are needed for making decisions? With the advent of digitalization and big data, do managers have perfect information? Making a decision is easy. However, making the right decision requires insight and knowledge. Knowing the weaknesses in information is critical for making wise decisions. Digitalization and big data do not directly translate to quality decisions.

Opportunities and problems compel managers to make decisions. A *problem* arises when performance does not meet expectations. Say, sales are lower than planned for a specific period. The manager needs to investigate why. Another example could be that employee dissatisfaction is rising; a third example might be that raw material deliveries are often delayed. An *opportunity* arises when circumstances unveil possibilities to improve results. An example could be the rising awareness of how food consumption impacts global warming, or the rising awareness of how food additives affect humans. Many consumers are shifting toward organic foods, vegetarianism, and veganism. While this could present problems for some food producers, food consumption itself is not dropping, so for others it may be an opportunity.

J. E. Sallis et al., *Research Methods and Data Analysis for Business Decisions*,
Classroom Companion: Business, https://doi.org/10.1007/978-3-030-84421-9_2

2.2 Problems, Opportunities, and Symptoms

When opportunities arise, the pressure to act is lower than when problems arise. Consequently, many opportunities are lost. Several opportunities are lost because companies do not systematically monitor their environment seeking them. Problems, on the other hand, will eventually show up in performance measures. If you could avoid problems, they would not be problems.

It is important to distinguish between *symptoms* and underlying problems or opportunities. A decrease in sales, though perhaps problematic, is a symptom of something else. Similarly, an increase in employee absenteeism or an increase in accounts receivable may be a symptom of something being wrong. Such symptoms can have many different explanations. Investigating why these conditions or symptoms are occurring is called *problem formulation* or diagnosis.

Equinor (formerly Statoil), the Norwegian state-owned energy company, provides us with an example of problem formulation. They conduct regular employee surveys measuring confidence in the corporate management. One year, the confidence level substantially fell compared to previous years. This outcome in itself gives no explanation or understanding of what happened (the phenomenon). However, it could trigger further research. If the subsequent research question were to "*understand* employee perceptions of corporate management," then in-depth interviews or group discussions (focus groups) would provide the answer. An alternative research question could be, "according to a set of predefined variables, *explain* the change in employee perceptions of corporate management." In this situation, an employee survey measuring the specific explanatory variables would provide the answer. In reality, Equinor assumed that the cause of falling confidence was due to unusual turbulence experienced in management that year and that the problem was therefore temporary.

In the Equinor example, confidence in corporate management had decreased, which was the symptom registered in their measurement. Their problem formulation was apparently quite simple. They attributed the symptom to management turbulence. Other cases may require a more systematic approach using both qualitative and quantitative studies. In practice, management often makes diagnoses without conducting research. To draw an analogy with medicine, management might conclude that the company has a cold, and the best action is no action. They wait for the cold to go over. Unfortunately, the company might have pneumonia that will not go away without treatment.

After diagnosis, the next question is what should be done? We determine a set of actionable options, and then analyze the potential outcome of each option. Of course, this assumes that the problem is adequately defined. In other words, there are two main issues to address:

- *Problem formulation*: What is the explanation of what has happened, is happening, or will happen?
- *Actionable options*: What can be done?

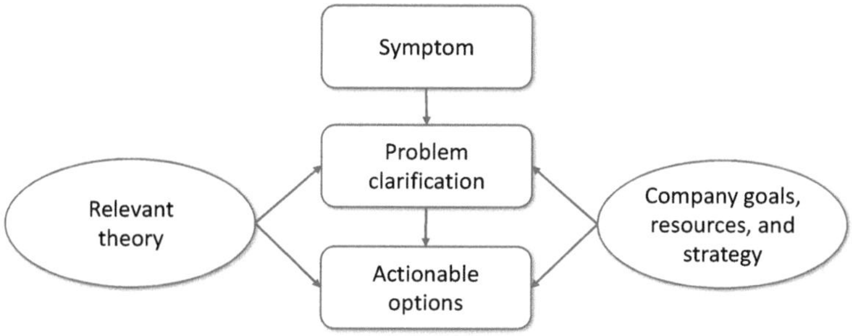

Fig. 2.1 Problem formulation and alternative actions

When it comes to both problem formulation and actionable options, it is advantageous to rely on relevant theory. A theory postulates which relationships exist between abstract phenomena (concepts) in given contexts. This means that a theory can help identify which factors can explain what is happening in a context, and can shed light on what actionable options should be contemplated.

Finding relevant theory is not always easy. First, many theories are relatively context specific in that they only apply to an isolated part of reality and assume that all other variables are constant. Second, theories sometimes only apply under certain conditions. Third, there may be different theories on what explains a particular phenomenon. For example, in terms of customer satisfaction, expectation confirmation theory postulates that a customer will be satisfied if the perceived value of a service is equal to the expected value of the service prior to purchase (disconfirmation paradigm). The absolute level of the perceived value of the service should, based on this theory, have no meaning for customer satisfaction. In simple terms, if you expect bad service and you get bad service, you are satisfied. An alternative theory postulates that the perceived value of a service is an independent explanatory variable that is of great importance to customer satisfaction. In simple terms, you are only satisfied if you get good service. Whatever the case, if customer satisfaction is low (the symptom), and if you know the relevant theories, you will be able to identify potential explanatory variables and how they are related.

The connection between symptoms and relevant theory is not always clear. Applying theory assume that the phenomena can be concisely broken down into clearly defined variables. If this is not the situation, then we have to clarify the problem without the help of theory. Managers just have to rely on experience, although a more robust alternative would be to use exploratory research to gain a better understanding for the symptom.

We visualize our explanation in Fig. 2.1. Relevant theory can help us clarify a problem by identifying which factors may be behind a symptom, and then the theory will indicate possible actionable options. It is important to note that the actionable options may not automatically follow the problem formulation. The company's goals, resources, and strategy dictate which options are viable. For example, it may be that the company does not have the resources to use a certain type of technology, although the problem formulation suggests that it is precisely the lack of this technology that is hurting sales. Perhaps the problem formulation indicates that it is web advertising that led to the competitor taking market share. If we do not have the resources to use web advertising, our options are limited to other actions.

2.3 The Research Purpose and Research Question(s)

It is always sensible to start by formulating the research purpose.. If the starting point is a symptom, the research purpose consists of two parts: One part clarifies what factors underlie the symptom, and the second part is to determine what action to take. For example, if we observe a decrease in sales, it is reasonable to first clarify why sales are declining and then consider options. If the starting point is an opportunity, for example, to launch a new product, the actionable options are usually given: launch or not to launch. Nevertheless, even when the options are apparent they may still require research, like clarifying the target segments in the product launch.

Often, the decision-maker is not the person responsible for the research. If the research is to be useful, it is crucial to have good communication between the decision-maker and the researcher. Otherwise, the decision-maker's response may be, "Well, that was interesting, but. . . ." Such a response is a clear indication that the problem definition was not good enough. It can often make sense to divide this process into two stages, and make sure that this happens in a collaboration between the researcher and decision-maker. The first thing is to formulate the *research purpose* in a single sentence. Then, formulate *research questions*, which together provide the answers we need in order to accomplish the research purpose.

Decision-makers will often have ideas and opinions about research problems. Anyone responsible for conducting research should respect this as a starting point. However, research questions of the type, "How can we achieve greater profits?," are too broad to be suitable for research. Nor should one accept a problem definition that already specifies the research design. The research design is determined after the problem is clarified, the purpose is defined, and research questions are specified. Here, we continue our analogy of the physician–patient interaction. Research shows that the alliance between a physician and patient is an important element of treatment and cure. Patients need to be heard and physicians need to listen. That said, it is not the patient who decides on the treatment—"My throat hurts, so please give me antibiotics." The physician, through a combination of patient dialogue, physical examination, tests, and expertise, diagnoses the patient and prescribes the proper treatment. In line with this, the researcher's role is to listen to the decision-maker, collaboratively arrive at a research purpose, conduct the research, and provide the relevant information. This is where the analogy ends. The information is a tool to assist the decision-maker; it is not the researcher who decides.

To clarify the problem, Chapman (1989) recommends that the researcher, based on a preliminary problem formulation, ask the decision-maker a number of questions. Examples of relevant questions may include:

- What decisions will be made based on this study?
- How will the results of the analysis influence these decisions?
- What information will affect the decision(s)?
- Why do you need to know this?
- What else do you need to know to make a sensible decision?

With answers to these questions, it will be easier to formulate a precise research purpose and relevant research questions. The following examples illustrate how to formulate the research purpose and research questions.

Example 2.1
A frozen pizza producer notes that their brand "Winter" shows reduced sales in the first half of the year compared to the first half of the year before. Management poses the question, why are sales dropping and what are our actionable options?

Decision Problem
What can the company do to increase the sales of "Winter" pizza?

Research Purpose
The purpose of this project is to identify which factors have contributed to the lower sales of "Winter" during the first half of this year compared to the first half-year last year.

Potential Research Questions
1. How has the market share for "Winter" changed from this year compared to last year?
2. Have new competitors entered the market?
3. How has the market presence of "Winter" (number of outlets/shelf share) changed since last year?
4. Have any specific customer segments changed their purchase frequency of "Winter" since last year?
5. How have brand perceptions of "Winter" changed since last year?
6. How does the price of "Winter" compare to competing brands?
7. How have consumer preferences changed toward frozen pizza since last year?

We see here that the purpose of the analysis is to identify which factors have led to the symptom we have observed (decrease in sales). When we formulate research questions, we build—directly or indirectly—on theory of what determines sales. The theory can help us get the best possible picture of what has happened, while the empirical survey will answer the research questions and lead us toward a solution.

An analyst purchasing standardized surveys prepared by market research agencies could answer many of the research questions in this example. Research questions 1, 2, 3, 4, and 6 can be answered with data from a retail category index.

Questions 5 and 7 could be answered from household panel data. These and other commercially available standardized surveys are discussed in Chap. 4.

If it had not been about groceries where data is regularly collected, new *primary* data would need to be gathered. This might especially be the situation for research questions 5 and 7. Properly answering them may require in-depth interviews or focus groups to formulate a questionnaire. Then, conduct a survey with structured questions measuring perceptions.

By answering, the research questions listed above, one can get an impression of what is probably behind the observed decline in sales. Thus, one will also clarify what kind of decision is most relevant to correct the development. For example, if we find that "Winter" has become less competitive in terms of price because competing brands have been offering discounts or systematically dropping prices, we need to focus on what we do around pricing. If the brand has reduced its representation at the retail level, this would be where we concentrate our energy. Of course, it is also possible that several factors have worked together and led to the observed "symptom."

Example 2.2

An oil drilling equipment manufacturer believes that it has developed a concept that is superior to existing solutions in a particular area. The question is whether to proceed with developing and launching this product. If so, how do they proceed to bring the product to market?

Decision Problem

Should the company launch the oil drilling equipment? And if so, how to proceed?

Research Purpose

The purpose of this project is to map out whether buyers are considering the planned product as a clear improvement in relation to existing equipment, and to uncover any barriers against them acquiring the product.

Potential Research Questions

1. How big is the total market in the relevant product area in different parts of the world?
2. Which companies buy the product and how high is their demand?
3. Which other manufacturers exist today and what are their market shares?
4. How do buyers view the various offers that exist today (both product and service)?
5. How do buyers view the planned product in relation to existing offers?
6. What is the purchasing process in this product area?
7. Are there different types of barriers that make it difficult to enter with a new product?

In this example, it is not as easy to find standardized surveys that the researcher can use as in example 2.1. However, within many industries, there are consulting firms that specialize in following developments within a particular industry. This also applies to the oil industry. Such reports may be relevant for an overview of the market structure. Publicly available reports, Norwegian and foreign, may be a useful source for answering research questions 1, 2, and 3. However, the more detailed information needed, the less likely it is that public information (*secondary* data) can be used. The researcher must then collect new *primary* data to answer the questions. In this case, it is reasonable to conduct a series of in-depth interviews with representatives of the purchasing department of the companies purchasing the equipment for research questions 4, 5, 6, and 7.

2.4 The Research Process

Figure 2.2 is a diagram of the research process. We take a pragmatic approach and frame the process around one or more decisions. We begin with a research purpose, and then formulate specific research questions. As we have discussed in this chapter, this forms the foundation of the research process.

When the problem is clarified, the research purpose is defined, and research questions are specified, we move on to the research design. It specifies the plan for conducting the appropriate type of research: things like what types of data we need in order to answer the research questions, how to obtain the data, and what type of

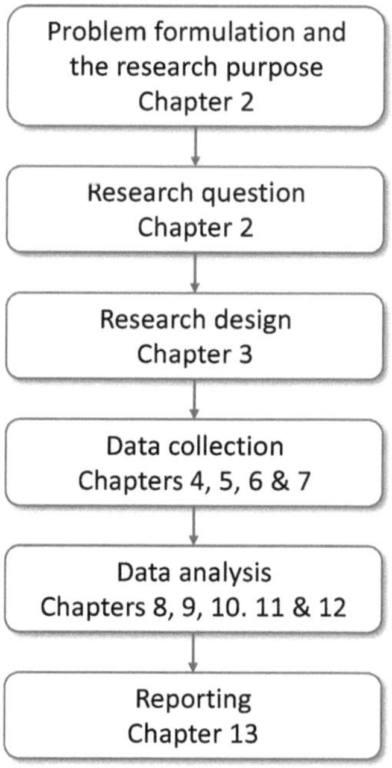

Fig. 2.2 Stages in the research process

analyses are appropriate. The individual research questions may require different research designs. Exploratory, descriptive, and causal are the three main types of research design (explained in Chap. 3). After the design is settled, data needs to be collected (explained in Chaps. 4, 5, 6, and 7), and analyzed (explained in Chaps. 8, 9, 10, 11, and 12). Reporting the research is discussed in Chap. 13.

2.5 Summary

Problems and opportunities arise in the course of doing business. In this chapter, we introduced the research process and explained problem formulation. Distinguishing between symptoms and underlying problems is an essential step at the beginning of the research process. Without the distinction, there is a great risk of studying the wrong thing. We touched on the role of theory for identifying relevant variables and their relationships in particular contexts. In lieu of theory, the researcher relies on experience and exploratory research. At this stage, our goal is to formulate a research purpose, which is further refined into research questions. Within the context of the research process, this is a logical procedure based on scientific principles of investigation.

Reference

Chapman, R. G. (1989). Problem-definition in marketing research studies. *Journal of Consumer Marketing, 6*(2), 51–59.

Research Design

3

Contents

3.1 Introduction

Good market analysis is largely about empathy, often expressed as the ability to put yourself in another person's shoes. The challenge comes when the people being studied will not or cannot express their opinions, or they literally do not know what is good or bad for them. How, then, can we develop products, services, and companies that create value? Innovation across industries is a major trend in entrepreneurship. Perhaps one has to look to other markets to gain knowledge of one's own market.

J. E. Sallis et al., *Research Methods and Data Analysis for Business Decisions*, Classroom Companion: Business, https://doi.org/10.1007/978-3-030-84421-9_3

Fig. 3.1 Research design

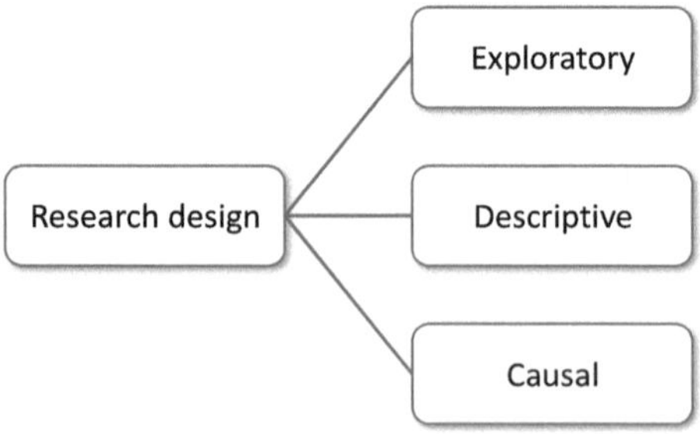

To begin with, research is about working systematically. Different approaches fulfill different analytical purposes for different research questions. The research design describes how the entire research process is set up in order to address the research question. Of particular importance are what types of data are needed, how to obtain the data, and how to analyze it. In other words, the design covers all the stages of the research process after defining the purpose and research questions. A fitting analogy is the architect's drawings and specifications for how a building is to be constructed. The choice of design depends on how much we know about an area and what ambitions we have in terms of analyzing and explaining relationships. From a pragmatic approach, it is common to distinguish between three main types of design: *exploratory, descriptive,* and *causal* (see Fig. 3.1). In this chapter, we describe all three. We will discuss what determines which design to use, with particular emphasis on *validity* and *reliability*.

3.2 Exploratory Design

If a decision-maker knows very little about a subject area, the primary goal of the research can be to explore the topic in more detail. An *exploratory research design* is appropriate when there is no clear idea of what variables are relevant or how they are connected. You may not know the relevant theoretical concepts or have a theoretical model as a starting point. In such a situation, the purpose is to gain insights into the research problem. The goal of the research can initially be to understand and interpret the relevant phenomenon in the best possible way. In many cases, it will also be relevant to develop hypotheses about possible relationships. A natural start in an exploratory design is to investigate whether something has been written about the topic (literature review), and whether there is data collected by others (secondary data). Often, it will also be relevant to collect your own primary data.

Imagine that you are facing a decision to change the wage system in a large organization. One alternative is to introduce bonus schemes. Theories of factors affecting employee job satisfaction may provide some clues; however, the theory alone is unlikely to provide clear guidance. An exploratory design, where the goal is to explore the employees' views on pay and working conditions, can be a good way to start, for example, by collecting data using in-depth interviews with a selection of people representing different functions within the organization. Then, combine this with secondary data on absenteeism and turnover in different departments. Very often, the purpose of an exploratory design is not only to understand the current situation, but also to develop hypotheses. Your in-depth interviews may indicate that motivation is particularly low in some departments, providing the basis to hypothesize that this exists for all departments.

In this book, we go through two main techniques for data collection in an exploratory design: focus groups and in-depth interviews. We start by describing the different techniques, and then elaborate on their practical implementation.

3.2.1 Focus Groups

In focus groups, a moderator leads a discussion on a specific topic with a group of about 8–12 people. Participants are selected to represent the opinions of a larger target group. Focus groups are particularly useful when assessing and developing new ideas. For example, one wants to find out the following:

- How does our target group experience a particular problem?
- What solutions do they prefer?
- What do they think of specific products or advertising messages?

Feasibility studies often use focus groups. Focus groups are useful for identifying key issues that are later investigated through, for example, data collected in structured interviews or online surveys. The group dynamic inspires participants to develop ideas and views. In this way, one can uncover conditions that are not so readily apparent in in-depth interviews or questionnaire surveys. The moderator is not interviewing the participants. The moderator's task is to keep the group on topic and ensure adequate discussion of the research problem. Both communication and observation are used. Based on the research questions, they manage the discussion and observe the respondent's nonverbal communication. Data analysis is done using qualitative interpretation and content analysis, possibly aided by using qualitative data analysis software like QDA Miner or NVivo.

3.2.2 Individual In-Depth Interviews

When the individual's personal experiences and opinions are of interest, individual in-depth interviews are most appropriate. With a prepared interview guide, the interviewer directs the discussion. The interview guide lists open questions that the respondent can freely answer. Individual interviews usually last for 1–2 h. In-depth interviews are also common for sensitive topics that are difficult to deal with in surveys or focus groups, for example, with sensitive topics that the respondent may not want to discuss with others present. There may also be practical reasons for choosing individual in-depth interviews. It can be difficult to gather a group of managers together in a focus group. Interviewer qualifications are similar to the moderator in a focus group. Interviewers and moderators must guide the interview or focus group without overly influencing the participant responses.

3.2.3 Other Techniques

In the exploratory design, *projective techniques* are an effective way of indirectly getting at a respondent's deeply rooted beliefs, attitudes, and opinions. They come from clinical psychology and consist of things like word association, sentence and story ending, and cartoon tests. The purpose is to get respondents to express their beliefs and attitudes in situations where they cannot or will not express them when directly asked. These may be sensitive issues or things that are difficult to verbalize.

To summarize, exploratory design is appropriate when little is known about a research problem. Literature studies and secondary data provide insights into the topic. In addition, it is very often necessary to collect primary data for a more detailed understanding. The most common data collection techniques are individual in-depth interviews and focus groups.

Exploratory Research Design

- Provides insight and understanding.
- Often uses literature studies and secondary data.
- Is a flexible process.
- Has two main techniques: focus groups and in-depth interviews.
- Is often used to generate hypotheses.

3.3 Descriptive Design

When you have a basic understanding of the problem area, a *descriptive research design* is appropriate for describing the situation in a particular context. This includes describing a single variable or the relationship between two or more variables. For example, in the aquaculture industry it could be important to understand the relationship between the quantity of harvested farmed salmon in different categories and the spot price level on the market. Based on economic theory, the hypothesis is that the price level decreases when the quantity offered increases. Using harvest data from salmon farms and prices from the spot market, you can analyze and describe the relationship between these variables. When you describe a context involving two or more variables, it is tempting to discuss them in terms of cause and effect. With a descriptive design, *we have no basis for claiming that there are causal relationships*. We can say that things change together in a systematic way, which we can refer to as correlation. However, we cannot claim that one thing causes the other. In our aquaculture example, we may find that harvest levels are systematically related to price. However, in our descriptive design, we do not say that price causes harvest levels or that harvest levels cause price.

Take note that in our aquaculture example we can use secondary data on harvest levels and prices to address our hypothesis. This is often the case with descriptive design. If we decide to delve deeper into the context, we might need exploratory or causal designs and we might need to collect primary data dedicated specifically to our investigation. This shows that design is a natural consequence of the research question and that data collection and the types of findings are a natural consequence of the design.

3.3.1 Survey Research with Questionnaires

The questionnaire is an instrument for collecting information that makes the communication between the interviewer and the respondents standardized (see Fig. 3.2). All respondents are asked the same questions in the same order, and they also get the same answer alternatives. An exception to this rule is when the respondent is asked to skip certain questions when they are not relevant to that specific respondent. If the answer is "no" to the question of whether you own a car, it is not relevant to ask questions about which car brand you have. Another exception is when an interviewer *probes* a respondent either to respond more extensively or at least give an answer. Strictly speaking, when interviewers probe respondents for answers, the personalized attention weakens standardization.

When doing questionnaire surveys, the researcher must carefully think about what questions to ask. As a rule of thumb, researchers should rely on already developed and validated questionnaires. When measuring distance, you do not invent your own measurement. You rely on standardized scales from, for example, the metric system. The same logic applies in social sciences. Many standardized measures exist. Look to them before creating your own questions. This increases the validity of the measurement (are you measuring the right thing), and the reliability of the measurement (are you measuring it consistently).

How the survey will take place is important. Is it a face-to-face interview setting or an anonymously completed questionnaire? The following techniques are common:

	Strongly disagree						Strongly agree
1. I trust that the other organization is able to fulfill contractual agreements.	1	2	3	4	5	6	7
2. We have made significant investments dedicated to this relationship.	1	2	3	4	5	6	7
3. There are many operating units involved from both organizations.	1	2	3	4	5	6	7

Fig. 3.2 Example from a questionnaire

- Web-based surveys.
- Postal surveys.
- Telephone surveys.
- Personal interview surveys.

The methods can also be combined, for example, by asking some questions by telephone and then sending a more comprehensive questionnaire in the mail to those who are willing to participate. Especially in research about companies where managers are the target respondents, telephone and then e-mail is common. When choosing, the individual methods have different strengths and weaknesses, which we return to later.

3.3.2 Observation and Diaries

Questionnaire surveys require that the respondents can adequately and honestly report on their behavior. When they cannot meaningfully do this, observation may be an alternative method for data collection (we discuss this in depth in Chap. 4). For example, we had master's students solving a case for a multinational cosmetics retailor. Their challenge was to propose how to enhance the customer experience with in-store displays. Customers are often not aware of their behaviors, and even if they are, they may not remember the details after the fact. The students, pretending to be regular customers, observed how the real customers interacted with the displays and behaved from when they entered the area to when they left it.

Another way to collect data is through diaries. Imagine you want to study television-viewing habits. Asking respondents to complete questionnaires about what programs they view and when is probably fraught with error. Do they remember everything they watched yesterday, last week, last month? Instead of a questionnaire, ask respondents to continually complete a diary about specific behaviors. Either during the behavior or at the end of each day, ask them to write down what they watched and when.

Smartphones and social media make registering (observing) behavior straightforward. This is why apps like Facebook are coming under increasing scrutiny and regulation. SoLoMo (Social, local, and mobile) is a mobile, local-based technology that is used to push advertising alerts on customers in certain geographical areas. These technologies record data such as mood, attitudes, and reaction patterns. Combined with geographic data, temperature data, and an infinite number of other data sources, these companies can compile an immense amount of detailed data about individuals.

Statistics Norway has used the diary method for many years when tracking people's time use (http://www.ssb.no/samfunnsspeilet/utg/200204/02). Time use tracking shows how the population distributes their time over different activities, and how the activities are organized in time and space. A representative sample, often referred to as a panel, of the population carry diaries where they record what they do and who they are with. They are also interviewed and answer questionnaires.

This is currently our most important source of knowledge about the unpaid work done in society. For example, the diaries tell who carries out household tasks, allowing researchers to calculate the amount and value of housework. They can compare this with labor force data to measure household productivity, or the balance between work time and leisure time.

Data collection using diaries is sometimes characterized as a form of observation; however, this is not really correct. After all, the respondents themselves register information in the diary. The only difference between this and a questionnaire is that registration in the diary takes place continuously.

Descriptive Research Design

- Can test hypotheses.
- Is often linked to quantitative analysis techniques.
- Is a formal and structured process.
- Often uses large, representative samples.
- Has three main techniques for data collection: questionnaires, observation, and diaries.

3.4 Causal Design

To investigate the possible effect of one variable on another, we use a *causal research design*. Generally, this means some form of experiment. To say that an event (X) is the cause of another event (Y) under a set of boundary conditions (Z), we must satisfy the following conditions that:

- There is covariation between X and Y.
- X happens before Y.
- And, other possible causes for the covariation are not present (isolation).

When scientists intervene in a situation to investigate a reaction, by definition, it is an experiment. In a simple experiment, we manipulate an independent variable to see if there is an effect on a dependent variable. In a social sciences experiment, participants are randomly assigned to one or more experimental groups, the independent variable is manipulated (called the treatment), and then the dependent variable is measured for changes. In business research, causality is often inferred from questionnaire surveys. However, given the dubious argument for satisfying the conditions of a true experiment, we call these quasi-experiments. Experiments are also classified by setting: lab experiments versus field experiments (see Fig. 3.3).

3.4.1 True Experiments (Lab or Field)

True experiments can be described by the following criteria:

Fig. 3.3 Experiment types

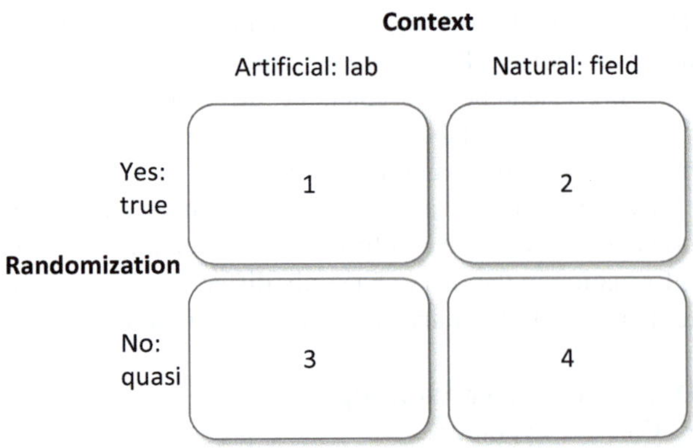

Table 3.1 True experiment

Time →			
	T_1	Stimuli	T_2
Group1: Control group	Pre-test	No treatment	Post-test Δ
Group 2: Treatment group	Pre-test	Treatment	Post-test Δ

- Random assignment of the participants into treatment and control groups (randomization).
- Pre-test (Time_0) both groups.
- Introduce stimulus to the treatment group.
- Post-test both groups (Time_1).

The measurement of each group's pre- and post-test levels on the dependent variable indicates whether the treatment (stimulus) has had any effect (see Table 3.1). Alternatively, only post-test can determine the level of effect, but then the evidence is weaker.

Marketing managers are often concerned with the effect of particular stimuli, like an advertisement, on some sort of outcome, like attention, interest, or purchase. Whenever the *effect* of something is in question, a causal design is appropriate for testing the relationship. We often do experiments to test the outcome of some action. For example, we raise the price of some product and observe how sales changes. Keep in mind that a true experiment requires that we systematically control for the effect of other variables that may influence sales. Imagine that we manipulate several independent variables and measure the effect on one or more dependent variables, while at the same time controlling for the effect of other variables. Suppose we want to show that lifestyle affects eating habits. Manipulating lifestyle to observe eating habits is nearly impossible. When it comes to stimuli such as advertising, it is quite possible to create experimental designs to investigate cause–effect relationships. This is done, for example, by so-called copy testing, where the effect of one or more advertisements is examined before a campaign is launched.

True Experiments Involve
- Randomization in experiment and control group(s).
- Manipulation of stimuli (treatments).
- Post-testing of the results in all groups.
- (Often) pretesting.

3.4.2 Quasi-Experiments

Quasi means almost. A *quasi-experiment* lacks at least one of the two characteristics of a true experiment, either *randomization* or *control group*. As a result, a quasi-experiment does not have the same rigor as a true experiment. Some of our students did a quasi-experiment about the effect of playing Mexican music on the sales of Mexican food. They could not randomize the experiment; however, they could have a control group. They chose two grocery stores from the same grocery chain. Both stores were about the same size and in the same town. They *pre-tested* by recording sales of Mexican food in both stores at the same time before playing the music. They played the music (the treatment) for 1 week in one of the stores, and then *post-tested* by recording sales at both stores directly after the music was played. They could then test for a significant rise in sales, and they could compare the results with the control group store. An additional effect was that all three students got recruited by a multinational consumer goods producer.

Quasi-Experiments
- Lack either randomization and/or control group.
- Have a weaker test of causality than a true experiment.

3.4.3 Lab Experiments

The Hawthorne experiment is classic in business research contexts. Lab experiments simulate reality in a lab setting. This is an effective way to have control over external environmental factors. In the Hawthorne experiment, to measure the effect on productivity, assembly workers performed their normal manual tasks in a special room designed to manipulate independent variables like light intensity, temperature, and break times. As the study progressed, productivity increased regardless of how the independent variables were manipulated. For example, productivity increased in both dim light and intense light conditions. The conclusion was that the worker's productivity did not change due to manipulating the independent variables. The change was attributed to the researchers paying attention to the worker's output. This is commonly referred to as the Hawthorne effect, otherwise called the observer effect.

We see from this example that one must always ask whether the results in an artificial lab situation are applicable in the real world. Participants can behave differently when comparing artificial and real contexts. Nevertheless, lab experiments can be useful, precisely because of being able to isolate external effects. Many retail chains, for example, use artificial stores to test the effect of different concepts.

Lab Experiments

- Take place in an artificially created situation.
- Make it possible to isolate the effect of stimuli because the environment can be controlled.
- May get results that do not generalize well to natural surroundings.

3.4.4 Field Experiments

A *field experiment* is done in the environment in which the phenomenon naturally occurs. As an example, imagine that a spare parts distributor for cars considers whether the company, as a policy, should give its customers gifts. To test the effect of such gifts, a random selection of customers is drawn and randomly divided into two groups. The *treatment group* gets a gift from the business while the *control group* does not get a gift. A questionnaire is then developed to collect data on customer's attitudes to five variables: price, service, quality, delivery, and loyalty. The results will show whether there is a significant difference between the treatment group and the control group. The field experiment thus provides information to the company with regard to the decision of whether to use such gifts on a larger scale or not. Field experiments can be real experiments with randomization of who gets the various stimuli, or quasi-experiments.

Field Experiments

- Are performed in the natural environment.
- Make it difficult to isolate the effects of our stimuli from other influences.
- Provide results that can be generalized to similar situations.

3.4.5 Internal and External Validity in Experiments

Internal validity is the degree to which the change in a dependent variable can be attributed to the change in an independent variable. For example, if we claim that X affects Y, we must be certain that X causes the change in Y, and that the change does not come from some other variable.

In order to isolate effects, that is, ensure high internal validity, it is advantageous to conduct a lab experiment. Then, only the stimuli being tested are manipulated. In a field situation, this is not as easy. Field experiments, therefore, have lower internal validity.

External validity is the extent to which research results can be generalized to similar situations. The external validity of field experiments is often higher than lab experiments because they are conducted in the real setting.

3.5 Choice of Research Design

There are three meaningful factors for choosing the research design: (a) experience with the subject matter, (b) knowledge of theoretical studies that identify relevant variables, and (c) the level of ambition with regard to identifying relationships between variables (see Fig. 3.4).

Having little experience with a problem means that we are primarily seeking to understand it, suggesting an exploratory research design. After the exploratory study, it may be appropriate to proceed with a descriptive or causal research design. This will depend on whether we succeeded in identifying variables and generating some hypotheses about their relationships. In theoretical studies of the context, they will usually provide the basis for identifying relevant variables and hypotheses.

3.5.1 Using Theory

Theories are useful to identify what variables and relationships are relevant for particular research questions. For example, Greve's (2003) article on R&D shows that by looking at the development stage and the decision stage in context, one can better predict why a company's degree of innovation varies over time, as well as explain the variation between companies. This contributes to explaining why some companies fail to launch innovations that they have developed. To be able to predict these outcomes, Greve identified a number of important factors for innovation: aspiration level, search, decision, risk tolerance, environment, and slack.

The example shows how theory can identify relevant variables. Therefore, an important task when conducting research is to identify what research has already been done within the research context.

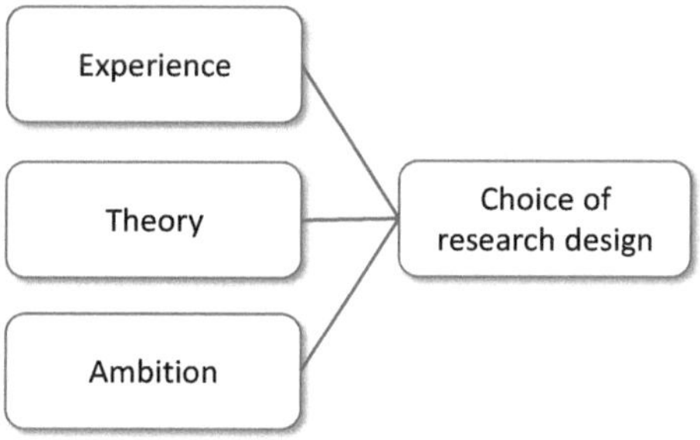

Fig. 3.4 Choosing a research design

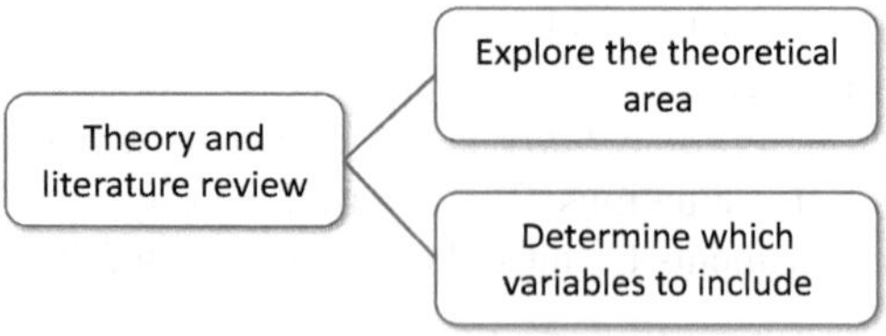

Fig. 3.5 Theory and literature review

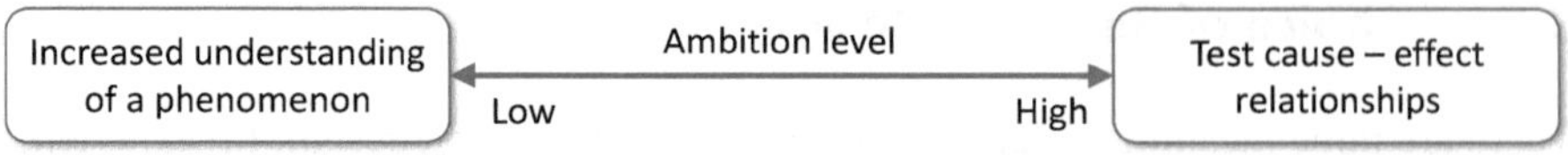

Fig. 3.6 Research ambition level

A literature review can have two purposes (see Fig. 3.5). If the research problem is unclear, we can use it to gain more knowledge about the theoretical area or context. If the research problem is clear, we can use a literature review to find out which variables are relevant and how they may be interrelated. The level of ambition in the research can vary from exploring a phenomenon to a detailed mapping of causal relationships between variables. An intermediate ambition can be to describe the covariation between variables (see Fig. 3.6).

The theory and literature review are used for two reasons:

- To increase the understanding and knowledge of the research topic (exploratory design).
- To understand which variables should be included and their relationship (descriptive and causal design).

3.6 Validity and Reliability

Whether planning, conducting, or evaluating research, it is evaluated by how "good" the process is carried out. In research, good is equated with validity and reliability. These terms are often applied to how well a phenomenon is measured. When we measure something, we use some kind of measurement instrument. This can be, for example, a questionnaire, an interview guide, a diary, or a technology that records behavior, like a mobile phone.

Validity refers to how well we measure what we intend to measure. *Reliability* refers to the consistency of how well we measure something. Two additional important aspects are random error and systematic error. *Random error* does not have any systematic source and has no pattern, so it tends to average out over repeated measures. *Systematic error* is consistent in each measurement, so over repeated measurements it causes a bias in the results.

The classic metaphor is to imagine a target. The very center of the target represents exactly what we intend to measure. Starting from the left in Fig. 3.7, our measures are spread randomly on and around the target so they are neither

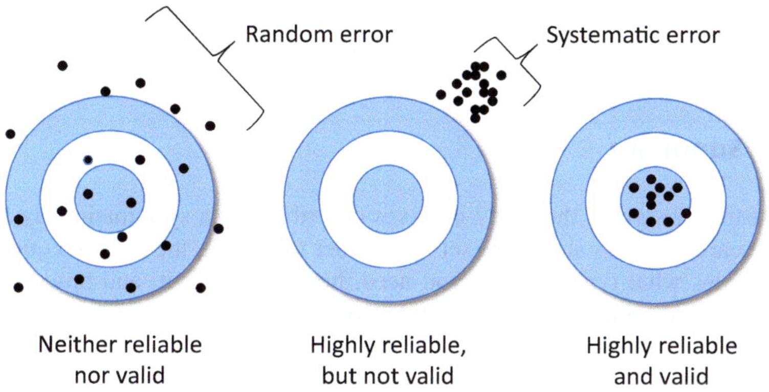

Fig. 3.7 Validity and reliability

reliable nor valid. They appear to have a lot of random error. On the middle target, the measures are consistently above and to the right of the target. We can say that the measures are highly reliable, but they are not valid. There is a systematic error that means we consistently measured the wrong thing. On the target to the right, we consistently measured what we meant to measure. We can say that our measures are highly reliable and valid.

Netflix provides an example of an objective measure. They track viewing habits like what viewers watch, when, and for how long. From the viewing profile data, they make program suggestions and most likely use it for internal decisions. However, in a household with two adults and two children, where two of them are male and two are female, results can be erroneous. With many family accounts, the users do not create individual user profiles. The "user" they identify is somewhere between 2 and 65 years old, is androgynous, and seems to binge watch, viewing several programs at once on different devices.

Now, it is important to make it clear that what you really validate is not a specific measurement method or test (although you can), but an interpretation of the data that emerges using a particular procedure. The interpretation may have a high degree of validity for one purpose, and a lower degree of validity for other purposes. In the Netflix example, the data may be reasonably valid for measuring family viewing habits. However, it is not valid on the individual level. Another possibility is that many people have their television on for company, yet their attention to the programming may be very fragmented. Objectively measuring "TV on time" for this group says very little about their actual viewing preferences.

It happens that researchers replicate other research, or possibly their own research in a new context. A growing trend stemming from medicine is called *evidence-based research*. It centers on doing research in a systematic and transparent way so that others can follow, replicate, and expand on previous findings. An absolute requirement of conducting good research, and a simple way to approach writing a methodology section in a report or book, is to explain what you did, how you did it, and why. Start with this in mind, and then frame it in terms of validity and reliability. If you do this well, then you meet the requirement for *replicability*.

3.7 Summary

The research design of the study involves describing in a systematic way how the entire research process should be set up in order to address the research question(s). This includes what types of data you need, how to collect it, and how to analyze it. We commonly refer to three main types of research design:

- Exploratory design.
- Descriptive design.
- Causal (cause–effect) design.

Theory and the literature review are used for two reasons:

- To gain increased knowledge and understanding of the research topic, often using an exploratory design.
- To understand which variables should be included in the study (descriptive and causal design), and how they may be related.

Whether planning or evaluating research, it is important to evaluate it in terms of reliability (how consistently something is measured) and validity (whether the correct thing is measured). An important aspect of this is replicability. We introduced two types of error:

- Systematic error (results in systematic bias).
- Random error (results in inaccuracy).

Reference

Greve, H. R. (2003). A behavioral theory of R&D expenditures and innovations: Evidence from shipbuilding. *The Academy of Management Journal, 46*(6), 685–702.

Part II

Data Collection

Secondary Data and Observation

4

Contents

4.1 Introduction

In this chapter, we discuss secondary data and collecting data by observation. *Secondary data* already exists, having been collected by someone else for another purpose. In contrast, *primary data* is collected to address specific research questions. Chapters 5 and 6 describe primary data collection. *Observation* is a structured way of collecting data about someone or something without direct communication. Observation may be collected as primary or secondary data. However, with the advent of the Internet and digitalization, the amount of secondary observational data being collected is staggering. It is revolutionizing data collection and analysis.

J. E. Sallis et al., *Research Methods and Data Analysis for Business Decisions*,
Classroom Companion: Business, https://doi.org/10.1007/978-3-030-84421-9_4

4.2 Main Types of Secondary Data

Figure 4.1 shows the distinction between primary and secondary data, and then between secondary data that is internal to the company being researched and data coming from external sources.

Big data is being collected by many companies without a specific research purpose, so it is secondary data. They collect it because they can. At the same time, many third parties, like Google and Facebook, collect digital data relevant to specific companies. For example, how many webpage views or how many likes a webpage receives. Then, there are many other sources of secondary data.

Secondary data can be useful in all research designs. The decisive factor is how well the data is suited to answering the research questions. If, for example, we want to investigate price developments in the farmed salmon market, we can look at secondary data from public sources.

4.3 Internal and External Sources

Sales, costs, revenues, and correspondence with suppliers and customers are all examples of internal data. *Internal data* is any source of information found within the company or organization that is not readily available to outside actors. The advantage of internal data is that it can be confidential, and thus not accessible to competitors.

External data is any source of information that is not proprietary to the company. This may be free publicly available data or it may come from subscription-based data suppliers like Euromonitor's Passport database. On their webpage, they describe Passport as a database providing insight on industries, economies, and consumers

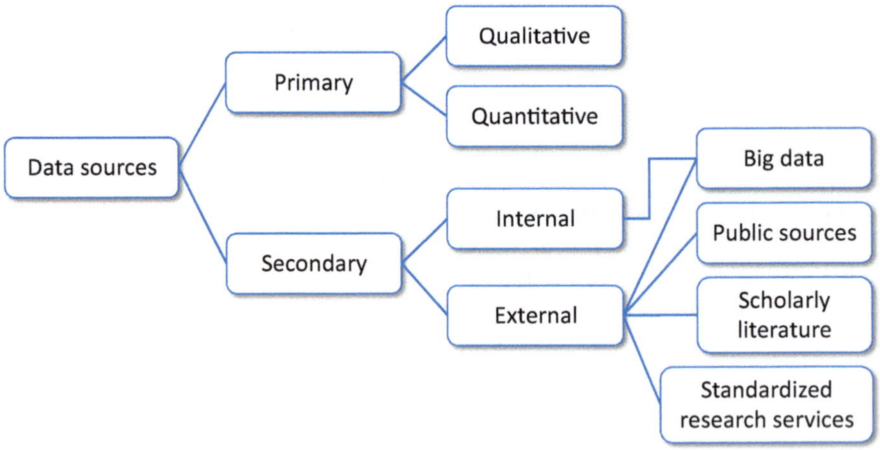

Fig. 4.1 Data sources

worldwide. It provides contextual data and identifies trends. Among the external data sources, we distinguish between big data, public sources, scholarly literature, and standardized sources like consumer panels run by data suppliers.

4.3.1 Big Data

Big data is secondary data that can be both internal and external. In just a few years, big data has gone from being a trendy buzzword to a phenomenon that businesses cannot ignore. Historically, communication has been the dominant way to collect business data, although big data is changing research in ways that we have yet to understand. While there is no widely accepted definition, *big data* refers to large amounts of data, so large and complex that traditional data analysis methods are overwhelmed. Arguably, big data may change the way scientific research is conducted. Mayer-Schönberger and Cukier argue that causal research will decline, and instead, correlations between variables will dominate since "Big data is about what, not why (2014, p. 14)."

The COVID-19 pandemic is a brilliant example of public authorities using big data to understand how it spreads, and thereby predict where it will spread and what to expect. In Scandinavia, the mobile telephone operators provided anonymized data to the governments so that they could track the movement of people within the respective countries. They could use the data to predict outbreaks and assign resources. In China, the government used big data to an even greater extent. The Chinese government's ability to observe individuals is a widely debated topic these days. For the pandemic, it proved quite useful. In train stations they complemented facial recognition with thermal imaging to identify potentially infected individuals, who could then be intercepted by healthcare personnel. The facial recognition technology also facilitated controlling whether individuals respected the quarantine. Mobile data was used to track their movements. The authorities launched the *Close Contact Detector App* to inform anyone who downloaded it whether they had been in the vicinity of an infected person. Similar apps were developed in Europe.

During the 2009 flu pandemic, the US Department of Health was keen to follow the spread of H1N1 virus, commonly known as the Swine Flu. Under normal circumstances, using data on registered cases reported by physicians, the authorities created a chart of the geographic spread of the flu. However, patients are usually sick for a few days before seeking medical help and physicians reported weekly on instances of the flu. Together, this meant that results were about 2 weeks behind the actual spread. Also in 2009, Google published an article in *Nature* reporting on how the company found a connection between the geographical spread of the flu in 2007 and 2008, by examining search terms used on the Google search engine. Their idea was that people used the Internet to obtain information as soon as they felt sick. Based on 50 million of the most common search terms and 450 million different mathematical models, they created a list of 45 search terms in one mathematical model that strongly correlated with the spread of the flu in 2007 and 2008. Using this, they could track the spread of the flu faster than ever before, almost in real time, and they could predict further spread (Mayer-Schönberger & Cukier, 2014).

Collecting huge amounts of data, what we now call big data, is closely linked to digitalization and the Internet, which started in the late 1990s. Before digitalization, countries collected large amounts of data every decade or so when they did census studies on their populations. It took massive resources and a lot of time, and the number of variables was very limited and they could not be updated in between the census years.

The ability to update and expand a database is a typical feature of big data. Results are available in real time. A well-known example is the technique pioneered by Amazon in the late 1990s to recommend books to customers. Ever since Amazon started its online bookstore, they collected data on customer purchases, what customers browsed before buying, which books were bought together, and many other variables. Initially, they used traditional customer segmentation; however, this led to recommendations of "more of the same." That is, if a customer bought a book on fly-fishing, then Amazon recommended more books on fly-fishing. Their breakthrough came when they realized it was unnecessary to use customer segmentation. Instead, they concentrated on finding the patterns (correlations) between books being purchased together, without caring about why they were bought together. This way, they could use all the data in the database and results were immediately available with each new customer visit. Often, books that were purchased together were quite different; however, that had no bearing on the recommendations. When Amazon broadened its offering to include other product categories, they applied the same method across categories.

In 2001, the consulting company Gartner described big data with three "V" characteristics:

1. High *Volume*: the amount of data.
2. High *Variety*: the variation in both structured and unstructured data.
3. High *Velocity*: how quickly is the data acquired.

Big data definitions often refer to data that by virtue of its volume is too complex (has too much variety) to be analyzed and treated by traditional data-processing software. Since the digital revolution, the real-time velocity aspect is included. An important characteristic to recognize is that big data is not usually generated for a particular purpose. It comes from many sources, which include:

- In-store transactions,
- Social media,
- Mobile phones,
- Sensors (built into goods or surroundings),
- Video surveillance (CCTV),
- Internet navigation (cookies),
- Online banking,

- E-mail,
- The "Internet of Things," like in-home online appliances.

Data occurs in two main formats: structured and unstructured. *Structured data* is what we commonly think of as organized, formatted datasets or databases where the data is organized in a predetermined format. The bulk of big data is unstructured. *Unstructured data* is not organized in any predefined manner and comes in a variety of formats like words, numbers, dates, phone numbers, and so on. Special programs are required to sort unstructured data into retrievable meaningful information.

How widespread is big data in Scandinavia, and what is it used for? In 2015, the SAS Institute, Intel, and Norstat conducted a survey with IT managers in the largest Scandinavian companies. It included telephone interviews and online surveys. They got responses from over 300 companies, although there is no information about the number or composition of non-respondents. 85% said that there is an increasing need to collect and analyze more data. 63% recognize a need for storing unstructured big data. 92% believe that big data storage and analytics provide a strategic advantage. Attitudes vary across industries, with telecommunications, media, and insurance services holding the most positive attitudes toward big data.

The IT managers were also asked which areas primarily need big data. They widely agreed that better knowledge of customers, "customer intelligence," was the most important area, especially in retailing, banking, insurance, and telecommunications. About half reported that their current employer had the infrastructure to use big data, with Finnish companies being the most advanced. Roughly, 25% reported that their company was in the process of implementing big data solutions, whereas about 25% were not yet considering it or it was a long-term issue coming on the horizon. Our conclusion is that most large Nordic companies are implementing big data infrastructure and that big data will play an increasing role in decision-making. For an overview of big data for practitioners and scientists, visit the "insideBIGDATA" website.

4.3.2 Public Sources

Public sources are places where data is publicly available. For example, "data.norge.no" is a Norwegian register of published datasets, inspired by, "dat.gov," which is an American website for the US Government's open data. Governments collect a lot of data and a lot of research is funded by public money. The premise is that data is more valuable to society when it is openly shared. When the data collection is publicly funded, then the public should have the right to use the data.

Other public sources include the national statistics agencies that most countries have, like Statistics Norway (ssb.no), Statistics Sweden (scb.se), and Statistics Finland (stat.fi). They offer data at the national economic level such as population numbers, consumer price indexes, opinion polls, social indicators, income, and so on. For research data related to (but not exclusively to) business, there are the Swedish National Data Service (snd.gu.se) and the Norwegian Center for Research

Data (nsd.no). They have publicly available data, sometimes for a fee, and offer services around data collection, analysis, methods, and so on. Many universities collect datasets, such as at BI Norwegian Business School (bi.no/bizreview), and Google has a specific search engine dedicated to research contributions (scholar.google.com). Rather than attempt to provide an exhaustive list, our point is that with little exploration, there is a lot of publicly available data from legitimate sources.

4.3.3 Scholarly Literature

Scholarly literature refers to sources that, in general, have been peer-reviewed. This means that they have gone through some sort of review process to assure a certain level of quality. Nevertheless, it is still extremely important to consider the source. As with everything on the web, you need to think of where it is coming from. Libraries can help sort through a maze of sources. Librarians are skilled at verifying sources. Many libraries have targeted search engines for finding books, research reports, student work, journal articles, and factual information. Usually, the sources are to some extent verified by the library so the information is reliable. Google scholar searches for virtually all published research articles, including reliable and unreliable sources.

Currently, most legitimate published research is covered by copyright protection, requiring some kind of paid subscription to get access. This is changing. Arguably, the more research findings are shared, the more value they have. By sharing, more researchers can move the research frontier forward. Having to pay for scholarly articles hinders their spread. Open Access Publishing is a way for authors and researchers to make their publications immediately accessible for free. The entire academic publishing industry is going through an upheaval as the business model shifts from subscriptions to Open Access.

4.3.4 Standardized Research Services

Standardized research is most often carried out by professional research agencies. There are many private actors in the business research industry, most of whom cater to marketing research needs. The largest ones have rich websites that provide a lot of information. There are reports on everything from international farmed salmon markets to male grooming in Scandinavia, to tourism on Svalbard. The most prevalent standardized research is for consumer markets.

Euromonitor's Passport database (go.euromonitor.com/passport.html) is a leading European source for economic, industry, and consumer data and reports. The Norwegian Strategy and Analysis Association (analysen.no) conducts standardized data collection, and offers seminars and courses to its members. Business Sweden (business-sweden.se) supports domestic and international trade, in part by offering data services and reports to Swedish companies going abroad, or foreign companies looking to invest in Sweden.

Nielsen (Nielsen.com) is the largest consumer-oriented market research company in the world. What started as manual inventory counting in 1933 evolved with the electronic and digital revolutions. They are active in over 80 countries and measure variables in a variety of settings like supermarkets, hypermarkets, convenience stores, and kiosks. The Nielsen market report handbook provides information on over 200 product groups for category development, price development, trends, chain structure, and so on. Nielsen Scantrack provides insight into:

- The grocery market and the service trade market broken down into chains,
- Market development in turnover and volume,
- Brand distribution,
- Retail pricing,
- Which factors contribute to market development,
- Promotional activities between competitors.

Gallup (gallup.com) is well known for its Consumer and Media polls. For over 80 years, Gallup has been conducting public opinion polls, first in the USA, and then worldwide. They are famous for predicting (or not) election outcomes. The core of their research centers on media habits, which includes readership for newspapers and magazines, radio listening, TV viewing, Internet use, streaming, cinema attendance, direct advertising, and so on. They connect this with consumption and demographic data.

What Do We Mean by Standardization?

In our examples of standardized research, the highest degree of standardization is when an identical report is offered to all subscribers without any customization. From this extreme, subscribers may only purchase parts of the research with relevance to their specific needs, or they may pay for customized data collection. It is possible that there is only one subscriber for specific parts of the research. Nevertheless, we still call it standardized because the data collection is part of a standardized system where the data is automatically collected without a specified research question.

Each research agency has its own format (method), which they usually describe. Their research procedures are essentially standardized, even when they collect data for different customers in different contexts. It is important to emphasize that standardized procedures are often applicable across unique research settings.

Standardized Panel Data

Panel data is collected from the same group of respondents (same sample) twice or more. The panel can be individuals or groups, like households or companies. The distinguishing characteristic is that it involves repeated measures over time. There are many panel data services like Norstat (norstatpanel.com) and GfK Sweden (gfk.com/en-se/). Consumer panels provide an excellent example of this type of data. Respondents (who are usually compensated) record their consumption habits. This is matched with other information like demographic variables and behavior variables

Table 4.1 Consumption habits, demographics, and behaviors

Consumption habits	Demographics	Behaviors
TV time	Gender	Transportation
Computer time	Age	– Car
Food purchases	Profession	– Bicycle
Clothing purchases	Education	– Public
Online purchases	Earnings	Pet owner
Offline purchases	Family situation	Sport enthusiast

to provide a detailed picture of consumption (Table 4.1). For example, a consumption habit of making online purchases may be related to the different demographics and/or different behaviors. Presumably, younger people make more online purchases than older people. In turn, younger people may be more likely to use public transportation than to drive a car.

An extremely important requirement for the validity of panel data is that the panel mirrors the population it is meant to represent (see Chap. 7).

4.4 Sources of Error in Secondary Data

Secondary data has the same weaknesses as primary data, and as such, it is important to know what errors can occur and how they impact the research. The great advantage of secondary data is that it already exists. It saves time and is usually cheaper than collecting primary data. In most cases, validity is lower because it was collected for a different research purpose. For this reason, secondary data is often used to provide insight to the research problem during the exploratory phase of an investigation.

In some situations, secondary data strengthens validity by providing an objective perspective on the research question. For example, when investigating the performance implications of human resource practices inside companies, by collecting questionnaire data from employees inside the companies and then measuring company performance from publicly available secondary performance data, the performance measure is more objective, and arguably more valid, than self-reporting on performance. Any data, primary or secondary, must be assessed for validity and reliability for the research it is being used in. It is important to know how the data was collected, how the variables were defined, and so on.

Sources of Error for Secondary Data

- Sampling and non-sampling error.
- Errors that make the data invalid.
- Errors that reduce the reliability of the data.

Sampling error occurs when a sample does not properly represent the population from which it was taken. *Non-sampling* error comes from several sources, for example, not collecting data from all the respondents, using the same respondent more than once, imprecise questions, leading questions, inaccurate answers from respondents, interviewing the wrong person, incorrect coding, misinterpretation, and data manipulation. When errors are random, they average out across the results, and so long as there are not too many, their influence on the results is minimal. When errors are not random, they systematically bias the results, which can have serious implications. Errors can be difficult to detect and remedy, so the best approach is to meticulously plan data collection to avoid error. In circumstances where data quality cannot be assured, the data should be discarded.

4.5 Collecting Data by Observation

A typical distinction when collecting data is whether it is collected through communication or observation. We address communication in detail in Chaps. 5 and 6. *Observation* is a structured way of collecting data about someone or something without direct communication. Figure 4.2 shows some of the different ways data can be collected through communication and observation.

It is nearly impossible to observe perceptions and attitudes. Collecting information about an individual's knowledge of something, their perceptions and attitudes, requires communication. On the other hand, if the primary interest is in behavior, what the individual does, observation is probably the best method for collecting data. In behavioral research, communication answers *why* a subject behaves in a certain way, whereas observation answers *how* a subject actually behaves. This is an important distinction because many behaviors are unconscious, the subject may forget, or in many circumstances, people do not want to admit how they behaved.

If the behavior takes place in public, like in a mall or at a school, observation is simple. However, it is extremely important to respect the ethical and legal restrictions on collecting data, especially if the target group is unaware of their participation. The European General Data Protection Regulation (GDPR) is an example of the growing concern with collecting digital data.

4.5.1 Types of Observation

Observation can be used on its own or as a supplement to communication methods. For example, in focus groups, one often observes the behavior of the participants. We record measurements of phenomena without asking for verbal or written responses. Imagine that you ask a focus group to discuss a new product that you give them to interact with. Through observation, you could record their level of confusion and the time it takes for them to understand the product. In some cases, the respondents or target group know that they are being observed; other times they do not. We can distinguish between how we are collecting the data, that is, the administration method. Usually, we distinguish between personal observation (someone watching) and technical observation (a recording device) (Fig. 4.3).

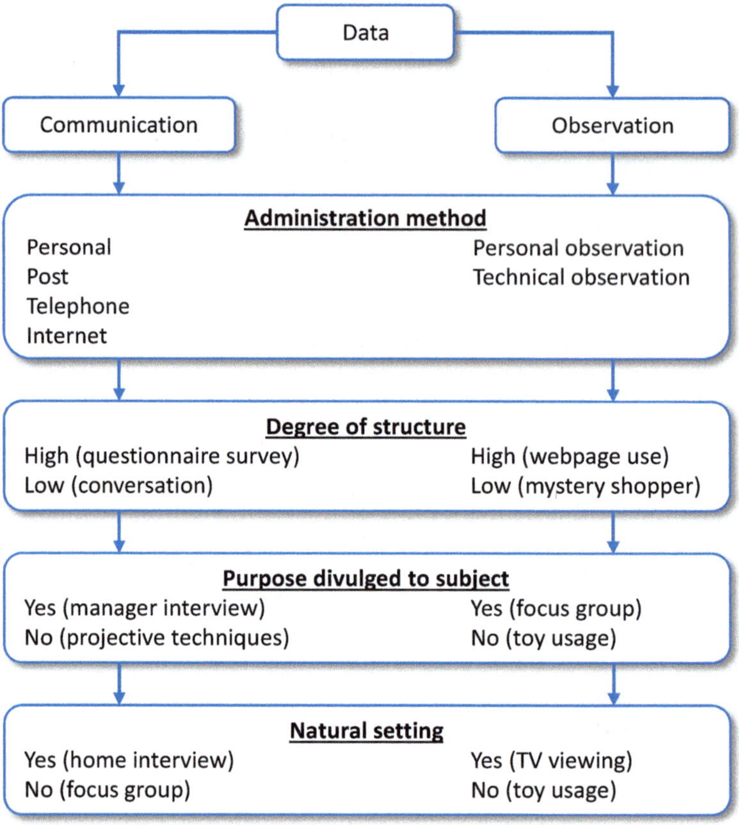

Fig. 4.2 Classification of data collection

Fig. 4.3 Types of observation

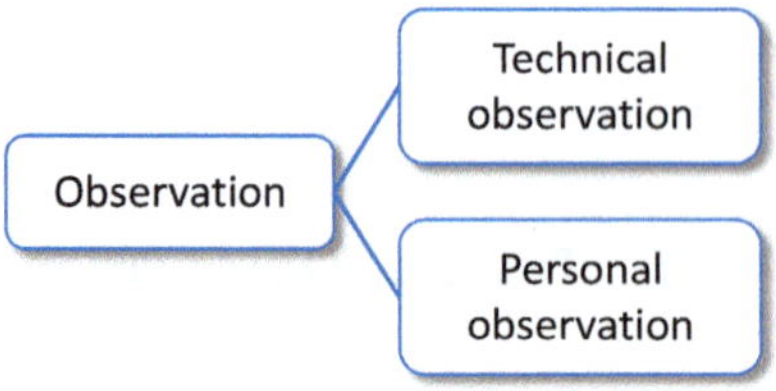

Tesla provides a good example of technical observation. They use mobile phone GPS data to monitor traffic density and movement. Based on phone movement and the concentration of phones in a geographic area, they can estimate traffic flow, which is fed to Google Maps and displayed on the dashboard navigation system. Algorithms enable data collection, processing, and reporting to take place in real time.

In recent years, television and radio broadcasting evolved from analogue to digital signals and many users shifted to streaming services over fiber or high-speed wireless. Rather than collecting viewing and listening data from questionnaires and diaries, content providers track usage by monitoring the digital signals. This improved accuracy, and thus the validity of the measurements. Of course, there is still no way to know the level of engagement on the user side. Is a person actually watching the TV, or is it just on for company?

The TV, radio, and Tesla examples are illustrations of *passive data collection*. Passive data collection is becoming more and more prevalent. It takes place without any interaction with the respondent, while capturing their preferences and behaviors, including location data from mobile devices, most often mobile phones. From the mid-1990s, HTTP cookies represent the first massive technical observation tracking system. Users were mostly unaware of or did not fully understand the degree to which their actions were being tracked. Tracking usage is often portrayed as a way to improve the user experience through customizing the interaction between the user and technology. Yesterday, my new telephone messaged me to say, "I noticed that you constantly unlock me at home. Would you like me to automatically disable the screen lock when we are at home?" While this is a great suggestion, it reminds me that my hardware supplier, the telephone operator, and numerous third-party apps know exactly where I am 24/7, assuming I carry my phone ☺. Legal regulation, like the European GDPR, is catching up and trying to assert some level of control over how much individuals can be passively observed.

In case you missed my "happy face humor," I did it on purpose to point out how often we transmit our mood, ☺☹, in our written communication. If you have Facebook, and you have provided your mobile number and permission to access your contacts, you probably also gave Facebook permission to mine your text messages. However, maybe you knew that because you read Facebook's terms and conditions—more humor.

Observation is also used in situations where respondents do not or cannot express themselves sincerely. *Mystery shoppers* are people who anonymously act as regular customers to record data on the in-store service they receive. Prior to the visit, they will determine what to record, for example, waiting time, friendliness, helpfulness, and product knowledge. From the customer perspective, the data will be much more valid with respect to true behavior than if the employees had been asked to self-report. Even if real shoppers were asked to report on the service afterward, they may have forgotten critical aspects or might feel reluctant to speak truthfully.

4.5.2 Measuring Emotions by Observation

Usually, specific types of behavior are measured by observation, either as they take place or afterward through behavior records. Background variables like age, gender, or place of residence are measured through communication. Admittedly, age and gender could possibly be assessed by observing a person, although accuracy, and thus validity, would improve through communication. Attitudes and perceptions are in many cases much easier to measure through communication, although there are examples of observations being used to measure emotional engagement (see

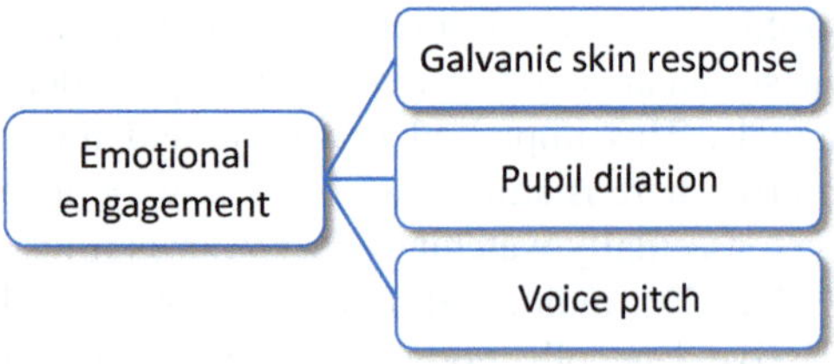

Fig. 4.4 Measuring emotional engagement by observation

Fig. 4.4). Galvanic skin response is a measure of electrical skin resistance, which indicates emotional stress. Eye pupil dilation and voice pitch analysis indicate level of attraction. Together, these observed measures could indicate emotional engagement.

Neuromarketing is the combination of neuroscience (study of the nervous system) with marketing. Neuromarketing uses various brain scanning equipment to measure how the brain responds to marketing stimuli like advertising and products. Often, brain responses are not consciously perceived, so this adds insight to other observational methods and reporting by the respondent. The theory is that emotional involvement and interest can cause physiological reactions that cannot otherwise be measured.

All of these methods have been used in studies of how people react to advertising and different package designs. Measuring voice pitch is the least invasive because we merely need to record the person answering questions. Measuring pupil dilation is more challenging, probably requiring considerable cooperation from the respondent. Measuring galvanic skin response literally requires hooking a machine to the respondent, and measuring brain waves is extremely expensive and invasive. Then, we have to make assumptions that the physical responses we see indicate specific emotional responses.

4.5.3 Using the Observation Method

In business research, observation is almost only used to measure behavior. In many situations, it provides more reliable and valid data than communication. One reason is that people simply do not accurately remember what they are asked about. For example, what brands they recently purchased or what music they streamed. Instead, these behaviors are more accurately recorded by registering brand purchases when paying or tracking from the source. Spotify's business model is based on tracking how many listens a song gets and then compensating the artist. At the same time, they track listener's history to, among other things, improve suggestions and customize play lists.

Another reason why communication does not always provide the best results is that respondents do not want to tell the truth. For example, if a gas station chain wants to evaluate how employees behave toward customers, they can of course ask the employees. However, they may receive more valid data by using mystery shoppers who act as regular customers and record the customer service experience.

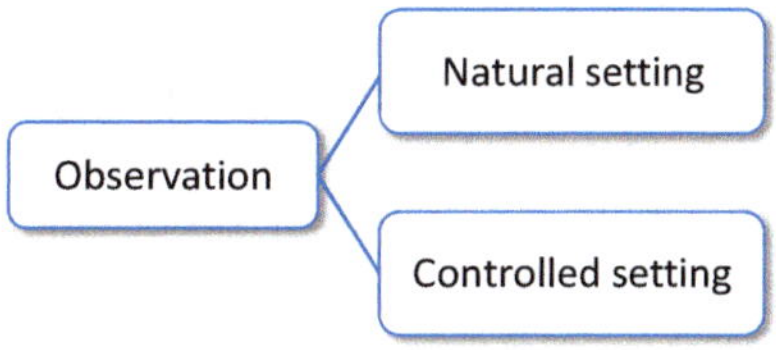

Fig. 4.5 Degree of structure in observation setting

Using mystery shoppers, though common, opens an ethical debate around data collection.

In some situations, respondents are not capable of reporting their behavior because it is unconscious or they lack the capacity to explain it. An example of the latter is when toy manufacturers test prototypes of new toys on children to learn how kids interact with the toy.

In the same way as with communication, observation can vary according to the *degree of structure* in the setting (see Fig. 4.5). If it is not predetermined precisely what is to be observed, then the degree of structure will likely be very low. For example, if we want to understand how children interact when playing, we may choose a natural setting, like a playground or kindergarten, to unobtrusively observe. If, instead, we want to observe their interaction with a specific toy, we might do the observation in a more controlled context at a market research facility. Controlled settings are more amenable to technical observation.

Some observations are extremely difficult to capture in natural settings. *Eye tracking*, meaning what we gaze at and how our eyes move, is a hot research area around our interaction with digital mobile devices, like smart phones and tablets. One area, which has been interesting to marketers for many years, is what we look at. What draws our attention and what are we interested in? Are we aware of banner ads? Given that we get a lot of information digitally, this has become relatively straightforward. By using software and infrared light technology, accuracy has vastly improved since the digital revolution. These eye tracking observations are certainly more reliable than if we asked what the person looked at. However, without communication, it is difficult to know what they are thinking. This is a weakness of observation.

4.6 Summary

Secondary data is a very effective source for getting information. It is subject to the same requirements as primary data for validity and reliability. While it may have been valid and reliable in its original context, this does not necessarily transfer to other settings. In organizations (e.g., businesses), we distinguish between internal and external sources. Big Data is both an internal and external source. Other external sources we discussed were public sources, scholarly literature, and standardized research services.

In the second part of the chapter, we discussed using observation methods to collect data. In business contexts, observation is most often used to identify or measure behavior. Compared with communication methods, it usually provides more valid and reliable data on actual behavior. Its weakness is that it is very difficult to understand why subjects behave how they do. We distinguished between personal and technical observation methods. With the advent of digitalization, technical observation is exploding into what is widely called, big data.

Reference

Mayer-Schönberger, V., & Cukier, K. (2014). *Big data: A revolution that will transform how we live, work, and think*. Mariner Books, Houghton Mifflin Harcourt.

Qualitative Methods 5

Contents

5.1 Introduction

Perhaps a little oversimplified, we can say that qualitative methods delve deeply into a subject to find whether something exists and to understand how it works. With quantitative methods, the focus is on measuring how much exists. Though the methods may seem incompatible. In fact, they are complementary and can be conceptualized as the two anchors on a scale. It is the research question that determines which method should be used in a given context, and in many situations, both are applicable. Using different methods along the qualitative spectrum on a specific research question is called method triangulation. It provides different perspectives on the same problem, and when the same conclusions are reached, it increases the validity of the findings. Applying qualitative and quantitative methods in harmony increases the robustness of the research. Both methods are used in social science research and business research because most social phenomena have both qualitative and quantitative aspects. Strictly speaking, it is the format of the data that determines whether it is characterized as either qualitative or quantitative. Quantitative data is expressed as numbers, while other data is referred to as qualitative.

Primary data can be collected in several ways, including:

J. E. Sallis et al., *Research Methods and Data Analysis for Business Decisions*, Classroom Companion: Business, https://doi.org/10.1007/978-3-030-84421-9_5

- Communication,
- Observation,
- Content analysis—which includes social media analysis and document analysis.

All three methods can provide both qualitative and quantitative data. It depends on how you proceed when you communicate, observe, or record the contents of social media and documents. When collecting qualitative data, we usually prepare a structured interview guide for communication or an observation guide to provide structure to collecting data on the phenomena we observe. The data is most often transcribed into documents that are then analyzed. Content analysis, of things like social media or documents, follows basically the same procedure as coding and analyzing transcripts from communication and observation. We explain this in more detail later in the chapter.

Flexibility is essential to qualitative data collection because it is an open interaction where we do not beforehand decide exactly what we will hear or observe. *Completeness* is to qualitative data what *accuracy* is to quantitative data. This is linked to the fact that qualitative analysis focuses on description and understanding of relationships, while quantitative analysis is usually to measure and generalize. The debate on the advantages and disadvantages of qualitative and quantitative methods in the social sciences is ongoing, possibly eternal. The important thing to always remember is that *the research question determines the method.* Somewhat simplistically, qualitative methods have their strength in explaining *what*, *why*, and *how*, while quantitative methods excel in explaining *how many*. Qualitative methods expose the *presence of something*, while quantitative methods show *how much*.

In business studies, in practice we often begin with a qualitative study (an exploratory design) before proceeding with a descriptive or causal design with, for example, a questionnaire survey. Often, a qualitative study will suffice. For example, if a series of in-depth interviews does not expose any interesting concepts or relationships in a specific theme, then there is little reason to carry out a quantitative study to measure something that was not qualitatively detected. Another example would be when you are trying to understand a phenomenon, like what people think about a product. Understanding will come from the qualitative research. If magnitude comes into question, then quantitative methods will come into play.

A qualitative study can also be carried out as a follow-up to a quantitative analysis. The purpose could be to get a deeper understanding for the quantitatively measured relationships between concepts. In one of our research studies on the relationship between financial advisors and customers, we were surprised to find that for advisors, getting the customer to trust them had a significantly negative effect on the advisor's understanding of the customer's risk preferences. We were left wondering, "How can trust in the financial advisor be negative?" Through interviews, we learned that when customers trust their financial advisor, they are happy to hand all responsibility over to the advisor. The advisors are happy because after they build trust, they do not have to spend much time meeting the customer (time is money). Unfortunately, the advisors understand that through trust, they lose contact with the customer, and the customer's situation evolves (they get married, have kids, change jobs), and thus their risk preferences change. Without qualitative follow-up, we would not have understood our quantitative results.

We will discuss focus groups, in-depth interviews, projective techniques, and content analysis of social media. These are but a few of the possible qualitative methods. For case studies, we suggest Yin (2009) as an excellent source for how to conduct case studies.

5.2 Focus Groups

A *focus group* is a deliberate gathering of people who discuss a predetermined topic with the aid of a moderator. They are often carried out by market research agencies or independent consultants on behalf of clients. The following situation often occurs at market research agencies. Eight to ten people are gathered around a conference table engaged in a discussion that lasts for a few hours. The group members (respondents) are not randomly selected. While they are usually not known to each other, they are usually selected to constitute a homogeneous group. The discussion is led by one or two group leaders (moderators) who are unknown to the group members. The interaction between the team members stimulates a discussion on predetermined themes. The discussion is recorded on audio and possibly video. In some cases, representatives for the client observe the group through a video feed. Based on the audio/video and the moderator's own notes and impressions, the content of the discussion is analyzed and interpreted. Usually, several focus groups are conducted for each theme.

The group dynamics are the focus group's foundation, and the purpose of using focus groups is to uncover conditions that individual interviews and surveys do not reveal. The philosophy behind focus groups is that "the whole is greater than the parts." The task of the moderator(s) is to ensure that the predetermined themes are adequately discussed. The moderator(s) will manage and redirect the discussion based on the purpose, as well as observe the participant's nonverbal communication. In other words, they use both communication and observation.

The data from the focus groups is processed, either through the use of qualitative interpretation or quantitative content analysis. The cost and time aspects are often highlighted as one of the benefits of the technique. Focus groups are used for different issues, and the implementation and data analysis are performed by people with varying qualifications. While focus groups may seem like a simple technique, there are actually several critical factors that require careful consideration before the desired information can be obtained.

Main Types of Focus Groups

In Calder's (1977) classic article, he distinguishes between three main types of focus groups based on the types of knowledge being sought (see Fig. 5.1). This classification has become the norm.

Exploratory approach: Used early in the research process to:

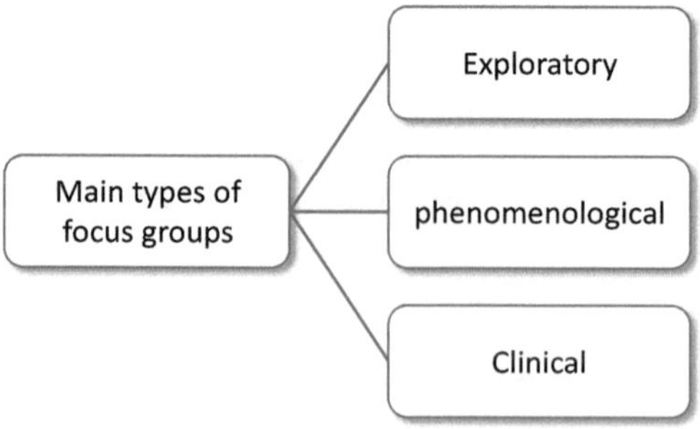

Fig. 5.1 Main types of focus groups

- Clarify and define research questions,
- Suggest hypotheses for testing,
- Develop new concepts (e.g., products or services),
- Obtain immediate reactions to new concepts,
- Pretest questionnaires,
- Interpret and understand results.

Phenomenological approach: Emphasis is placed on understanding the target group's everyday experiences and life situation to:

- Learn the "language" of the target group,
- Understand their situation and problems,
- Map needs and satisfaction.

Clinical approach: The purpose is to understand latent relationships to:

- Gain insights into motives that are difficult to get by using a structured questionnaire because the respondents themselves are not aware of them.

The following benefits and limitations are often mentioned when describing the focus group technique:

Benefits of Focus Groups

- Compared to individual interviews, focus groups generate data faster and at a lower cost. They are also quicker to implement when compared to many individual interviews.
- Focus groups enable direct interaction with the respondents, so topics or issues that come up can be immediately followed up for deeper insights. Nonverbal responses can be observed, which may provide important signals to the moderator and valuable data to the analyst.
- In focus groups, participants can express themselves in their own manner. Latent understandings and relationships can be uncovered.
- Group dynamics can stimulate participants to open up and provide additional information.

- Focus groups are flexible and can be used for different themes, with different participants and in varying environments.

Limitations of Focus Groups

- The low number of participants, their selection criteria, and the moderator as part of the measurement instrument prevent the generalization of results. That is, the participants may be different, and thus not representative of the population of interest.
- The "researcher" or the decision-maker places too much emphasis on the participant's statements, forgetting that the statements may not represent the population.
- Dominant people in the group can largely influence the results of the focus group. This prevents some respondents from expressing their opinion.
- The open discussion format in focus groups results in a lot of unstructured data being generated. This is challenging to analyze and interpret.
- Consciously or unconsciously, the moderator might influence the discussion to obtain desired results rather than valid opinions from the participants.
- Focus groups are difficult to replicate, making it difficult for subsequent researchers to precisely verify and validate results.
- There are few scientific studies of the focus group's strength compared to other qualitative methods and techniques.

5.3 Individual In-Depth Interviews

Individual in-depth interviews are conducted on a one-to-one basis, usually lasting about an hour. They are conducted when the individual's personal experiences and opinions are of interest. To saturate a topic, it often takes 15 to 20 interviews. They should be recorded and then transcribed for data analysis. All told, this is very resource-demanding and costly, especially if it includes travel. Compared to focus groups, the data probably comes from fewer people, although it is the quality of the data that is most important. It is the research question that drives methods choices, and some topics are more appropriately investigated through individual interviews than focus groups. The more sensitive the subject, the more logical it is to use individual interviews. Sensitive themes can also be addressed in focus groups. However, this assumes that the participants are confident in each other and have accepted to freely talk about a particular topic.

Compared to focus groups, the interviewer has more control over the direction of the conversation. The respondent has more time to express ideas and get direct feedback from the interviewer, although the interviewer has to be careful to allow the respondent to talk openly. This is why the interviewer usually uses an interview guide. It helps keep the interview on topic and makes sure all topics are sufficiently covered. Keep in mind, that just like focus groups, interview findings cannot be directly generalized to the population.

Benefits of Individual Interviews
- High degree of individualized information,
- Easier to discuss personal information,
- No group impact.

Limitations of Individual Interviews
- Resource demanding (costly),
- The interviewer can inhibit the respondent.

5.4 Projective Techniques

Projective techniques use stimuli, like visual aids or sentence completion, to indirectly evoke beliefs and attitudes. They are based on methods used in clinical psychology. The purpose is to get respondents to express their beliefs and attitudes in situations where they cannot or will not express them to direct questions. This may be due to the fact that it is a sensitive topic, or that it is difficult to express the views verbally. The most well-known example of using a projective technique is probably the Mason Haire (1950) study to reveal why instant coffee was initially unsuccessful in the US market. Traditional questionnaire surveys indicated that housewives did not purchase instant coffee because of the poor taste (at the time, housewives were overwhelmingly responsible for household grocery purchases). Taste tests, where respondents were asked to taste coffee prepared in different ways, however, did not indicate that the taste alone was a major problem. Haire wanted to find out if there were other reasons why housewives did not buy the instant coffee. He prepared two lists of seven goods that a housewife might have bought at a grocery store. They were identical except that on one list there was Nescafé Instant Coffee, and on the other, Maxwell House Coffee (regular roasted coffee).

Maxwell House was the leading US brand at the time. With two groups of 50 randomly selected housewives, one group was given the grocery list with the instant coffee, while the other group got the grocery list with the normal coffee. All 100 housewives were then asked to, in their own words, describe the housewife who bought the goods on the shopping list. When the housewives projected their attitudes into the description of the person who had bought the goods, they described the instant coffee purchaser as lazy and poor at planning purchases. They called the instant coffee person a spendthrift (cheap), whereas the regular coffee purchaser was resourceful and a good wife. Of course, the study reflects the attitudes of the era.

There are several advantages to using projective techniques. First, the respondents can express their feelings without putting them into words. Second, they can respond intuitively, without having to justify their response rationally. Third, many respondents like this sort of creative task more than answering traditional questions in surveys. This contributes to more attention from the respondent, which increases the validity of the answers and higher response rates.

Projective techniques can be divided into four categories:

- Associations (When you see ____, what word comes to mind?).
- Sentence and story-ending, metaphors, and analogies (My smartphone is a____).
- Third-person projections; what do other people think, feel, do? (My friends use ____ because ____).
- Expressive techniques; role-play, drawing, etc. (If you were the store manager, what would you do?).

The classic instant coffee study used third-person projection; however, associations, sentence completion, or expressive techniques could have all led to similar results.

Metaphors have been widely used to uncover the basic beliefs people have about an organization or product. For example, which animals are associated with different car brands. Personification is a special form of metaphor where respondents are asked to describe the sort of person who is associated with something. This is commonly used in branding. Hofstede et al. (2007) show how they used personification to distinguish the brand positions of four beer brands on the Dutch market. They used two variants: the first was based on associations between beer brands and celebrities, while the second was based on associations between beer brands and various professions. Below, we describe the first variant, since the other was done in almost the same way.

First, 16 participants were asked to individually go through a number of magazines and cut out pictures of celebrities who they felt fit with each of the four beer brands. These pictures were glued onto a page for each beer brand, which each participant had at their disposal. Subsequently, the participants were asked to pick out a personality trait that characterized each of the celebrities who they had selected based on an alphabetical list of 73 personality traits. The results showed interesting aspects of how the beer brands were perceived. For example, the respondents had great difficulty finding any celebrities whatsoever suited to beer brand B. This suggests a very unclear brand profile. In the case of beer brand A, there was a large predominance of Dutch celebrities. It was interpreted as a strong domestic beer brand. Beer brand C had many well-known pop and rock artists, while beer brand D had the largest range in celebrities, from members of the royal family to athletes. Brand C was interpreted as being the hip beer, while Brand D was interpreted as being well established and widely accepted by the population. Such interpretations are often called *holistic*, since they are based on a broad overall view of the results. When assigning personality traits to the brands, it was difficult to find a pattern between the celebrities and the professionals. This led the authors to question the validity of transferring results from projective techniques to individual variables.

5.5 Content Analysis of Social Media

Social media opens up a host of opportunities for conducting qualitative studies. Social media analysis is one aspect, while another is the dialogue within groups (discussion forums, blogs, and Internet communities) that were previously almost impossible to observe. Many companies organize their social media platforms to facilitate analysis. For example, Facebook has its own website that shows how to get analyses and insights from Facebook pages (developers.facebook.com), Google has analytical tools (analytics.google.com), as well as Adobe (adobe.com/analytics) and Twitter (analytics.twitter.com). The competition is intense and new players are constantly entering the market.

For companies that actively manage brands, monitoring social media is an important source of information on how attitudes and preferences toward their brand are evolving. For example, Blog Mining (mining.com) is a tool for monitoring thousands of blogs simultaneously. It is common to develop *Dashboards*, which are visual tools that present real-time information on things like key performance indicators and competitor benchmarking parameters. Their purpose is to, in a user-friendly way, present data that matters. Companies can participate in discussion forums to get perceptions and opinions about brand extensions, product features, packaging, and the like. Web scraping and web crawling refer to tools that actively scan and harvest data from the web. These are excellent tools for spotting trends.

Social media can also be used to develop communication with different target groups. Let us say you want to develop a health product. One way to get information on how the target audience associates with health is to explore *online message boards* and see what is tagged with "Health." Then, see what other keywords are used in the context. Perhaps you can even establish your own online message board and see how others react to it. In this way, you improve your interaction with the target group. Other sources include Instagram, which can be used to uncover associations and image usage, and LinkedIn, which can enrich the understanding of a phenomenon and the people who engage in it. With the help of *word cloud generators,* you can get the most important words associated with concepts, products, and brand names.

5.6 Problem Formulation and Qualitative Data Analysis

In this section, we discuss five generic stages in qualitative data collection, analysis, and reporting (Fig. 5.2).

(1) Problem Formulation

In stage 1, the researcher and the client (assuming there is one) need to agree on a common problem and research questions. If the research question is practical and exploratory, theory may not play any role. However, it is important to recognize the value of theory in understanding the phenomenon and indicating ways to explore it. Specifically, theory can indicate which variables are important, how they are related, and even ways they can be measured. This forms the foundation of the entire process. Important questions are:

1. What is the research purpose?
2. What types of data are required?

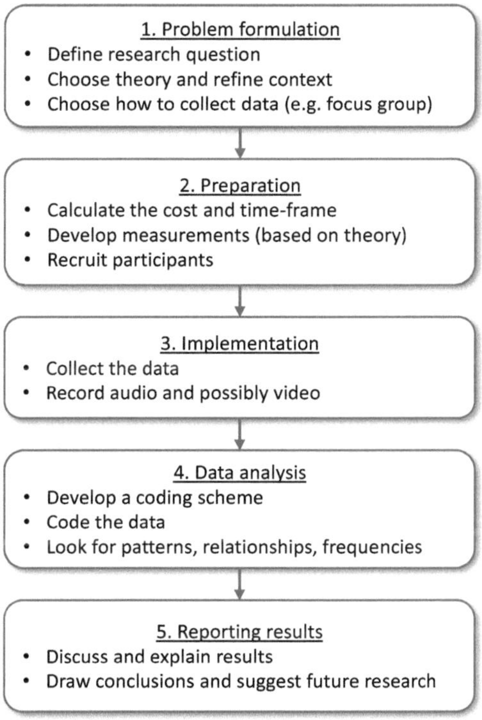

Fig. 5.2 Stages in conducting qualitative research

3. What knowledge do we have about the research area?
4. Who is the target group (primary, secondary)?
5. What will the results be used for?

Based on a common understanding of the problem, one has to decide on what type of data collection is appropriate, for example, interviews, focus groups, or text analysis. This also provides guidance for:

- How much data is needed? For example, the number of interviews or focus groups.
- The composition of participants (heterogeneous versus homogeneous).
- What sources of text are relevant, like which social media forums or which documents.

As guidance on how much data to collect (e.g., the number of interviews or focus groups, or the amount of text to analyze) you will discover that in the early stages of data collection you will be surprised by findings. As the surprises diminish, you will realize that the topic is saturated and there is nothing more to learn. It somewhat depends on the homogeneity of the topic. 10–15 interviews may suffice, or 3–4 focus groups. It is a judgment call with respect to when you have saturated the topic.

It is most common to have eight to ten participants for exploratory or phenomenological focus groups, and six to eight for clinical focus groups. Small groups can be less productive and more expensive than larger groups. Fewer than six respondents can lead to productivity issues because participants are more sensitive to group dynamics. With large focus groups, more than twelve participants, it is difficult for the moderator. One danger is that large groups break up into smaller groups that discuss the same subject. This complicates both the management of the focus group and the interpretation of the discussion.

Another important aspect is whether focus groups and interview participants should be homogeneous or heterogeneous. In exploratory research, heterogeneous participants provide rich information from broad perspectives. For focus groups, since the phenomenological approach is generally focused on learning about a target group, homogeneity is more likely to lead to consensus around a common perspective. The clinical approach has smaller focus groups with no specific requirement for homogeneity. If there are several distinct segments in the target group, there may be a need to use separate groups (e.g., only men or only women). The more homogeneous the participants, the fewer focus groups or interviews needed because they relatively quickly reach consensus and the topic becomes saturated.

(2) Preparation

This is an intermediate stage where the researcher organizes the practicalities of conducting the qualitative data collection. Five things are particularly important to consider:

1. Draft a discussion guide, interview guide, or set of search criteria for text analysis.
2. Create a participant profile if data is collected from people.
3. Calculate the cost and time.
4. Book facilities. The setting can greatly influence how participants engage.
5. Recruit participants.

The *discussion guide* or *interview guide* is a document the moderator uses to make sure that all relevant aspects of the discussion are covered. It consists of an overview of topics and questions to be addressed, and is usually on two to three A4 pages. While it offers a structure for how the discussion can unfold, it is not a questionnaire and should not hinder the natural development of the discussion. The moderator can use it to cue certain themes or topics that need discussion. After the first interviews or focus group, the guide may evolve to reflect topics that come from the respondents. For example, in about 1999, we were collecting data from retail grocery chains in France about their relationship to Norwegian salmon suppliers. In one of the first interviews a respondent said, I am terrified of *mad-fish disease*. As a result of the ongoing beef crisis with mad-cow disease, there were spillover effects throughout grocery supply chains. Traceability of product source had become very important. In subsequent interviews we always asked, "Are you concerned about mad-fish disease?" The question evoked excellent data on how retailor–supplier relationships were evolving.

The participant profile is a description of who should be recruited to participate in the interviews or focus groups. This is based on the profile of the target group (the population), and in turn, the degree of heterogeneity. Recruiting respondents is time-consuming and should be addressed early in the process. Potential participants must match the participant profile. The quality of the participants will greatly affect the results.

Costs come into play when using a professional moderator, technical equipment, renting somewhere to meet the participants, and so on. Participants may want to be paid or receive some sort of recognition for participating. As with many things, good planning reduces unexpected events that may negatively impact the research.

(3) Implementation

In contrast to quantitative research where the measurement instrument (e.g., questionnaire) is fixed, the measurement instrument in qualitative research is flexible. This important distinction allows for the moderator's subjective influence. The moderator acts as the link between the discussion guide and the respondents. The purpose of the interaction is to get the respondent's contribution, and the moderator's main task is to ensure that the topics are satisfactorily discussed. To assure this, the discussion between the respondents is held in line with the interview or discussion guide. If the guide is deficient, and/or the moderator lacks the necessary competence, the results may be less valid.

The duration of the focus groups or interviews is important. The group dynamics need time to develop in a focus group. As a rule, participants need some time to feel comfortable and they must be given sufficient time to express their views. If they feel time-pressure, it can create anxiety, resulting in incomplete and potentially invalid data.

Individual personalities and combinations of personalities can affect the behavior of individuals in focus groups as well as group dynamics. Group characteristics in this context include both sociodemographic variables (age, gender, income etc.) and psychological characteristics. Interaction in the group is more harmonic when the socioeconomic backgrounds are similar. A dominant and/or aggressive person can disturb the group dynamics and thereby reduce the focus group's quality. The moderator's role here is central, and a trained moderator will more easily manage such situations. The more knowledge the moderator has about different types of personalities, the better he/she is able to resolve conflicts. A good moderator will quickly assess the respondents.

The physical environment for conducting a focus group also affects the group dynamics. The room should be furnished so that the respondents can easily see each other and the moderator so that they can focus their attention on the discussion. For one-on-one interviews, the setting needs to be such that the respondent can relax and engage with the interviewer.

Observers may be researchers, or in a commercial context, the client. They may passively sit in the same room, or watch from an adjoining room through a mirrored window or video link. It is important to recognize that participants may alter their behavior as a result of being observed. In the same way as being recorded, participants will probably gradually forget that they are being observed.

(4) Qualitative Data Analysis

Rather than thinking of qualitative versus quantitative analysis, it may be easier to envision an inductive versus deductive approach. Strauss and Corbin (1990) proposed an inductive *grounded theory* approach, whereby the analyst interprets the data without any preconceived notions of concepts or relationships (no theory). Starting with an open mind, through applying a grounded theory analysis, the researcher works inductively toward forming ideas about the meaning of the data.

A deductive approach starts with a theory, which may be formalized or simply a preconceived idea of how things work, and uses the data to test the theory. We view it as a cyclical progression where, starting with the discovery of a phenomenon, the researcher inductively develops a theory about how it works, then deductively tests the theory, and then depending on the results, moves back toward discovering new ideas to revise the theory, and then tests again, and so on. You could call it, organized trial and error. We discussed this in Sect. 1.3 (see Fig. 1.3), when we explained our interpretation of falsificationism.

Not always, but most often, the qualitative data takes the form of a written transcript. Whether it is a transcript from focus groups, interviews, observations, or things like social media interactions, it needs to be coded. The process can be done by hand, or as we will demonstrate, with the support of coding software.

Example 5.1

Figure 5.3 shows a content analysis of an interview transcript of a manager who was asked, "Please describe and give examples of how you learn from the companies you partner with." The analysis was done in a qualitative data analysis software called QDA Lite. The analysis follows the basic routines of open coding where words or pieces of text are related to concepts. We imported the interview transcript and then prepared a set of codes that were relevant for understanding the topic, which in this case is inter-organizational learning:

- Promote (things that promote learning),
- Hinder (things that block learning),
- Incremental (meaning gradual learning),
- Discontinuous (meaning higher-level learning of radically new things),
- Cooperate (how cooperation influences learning),
- Compete (how competing influences learning).

We then highlighted pieces of text that were representative of each code. As we coded, we revised the coding scheme to better match our understanding of the text.

A challenge with qualitative data analysis is to ensure high qualitative rigor, while at the same time retaining the creativity to generate novel findings. The

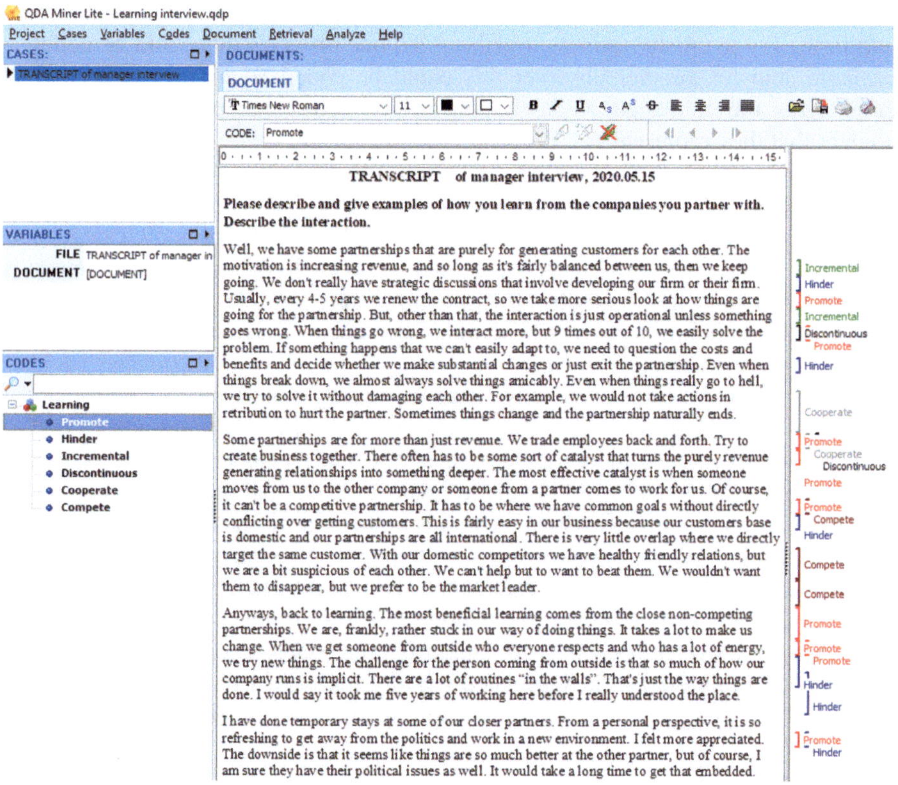

Fig. 5.3 Content analysis in QDA Lite

research should be grounded in a well-specified research question. From there, an appropriate research design that will provide the data to address the research question. In the organizational learning example, the transcript is the data and the method is content analysis. The research question could be: *What are the important factors that promote or hinder inter-organizational learning and is there any distinction between incremental and discontinuous learning?* ◀

The Coding Process

Following the process outlined in Gioia et al. (2013), the first step (*first-order analysis*) is to read the text and categorize based on the informant, or in this case, the manager's own words. Proponents of grounded theory would call this *open coding*. This evolves into creating a set of codes by which the researcher classifies words, sections, pictures, or whatever is in the content into categories. In the current example, we have a code called *promote* and another called *cooperate*. We differentiate between *promote*, which is anything in the text that may promote inter-organizational learning, and *cooperate*, which is a well-established construct that is known to promote learning. Deciding whether a phrase should be coded as *promote* or *cooperate* is difficult. Consider the phrase, "We trade employees back and forth. Try to create business together." To one observer this could be coded as *promote*, and for another observer this could be coded *cooperate*, or both. This emphasizes the challenge of developing robust codes that properly capture the content that they represent.

The next step (*second order analysis*) is to identify similarities and differences among codes, and then form categories of codes. In grounded theory, this is called *axial coding*. Perhaps cooperation should be merged with the promote category, or perhaps a new higher-order category should be created. When the data achieves a coherent structure, the researcher then identifies relationships between categories. This may be an iterative process benefiting from deeply understanding the data and contrasting it with existing theory.

Coding with software has the advantage of creating a hierarchy of codes that can be edited and organized as the analysis proceeds. It allows for automated coding based on word searches and phrase searches. The analyst can create outputs like bar charts, pie diagrams, and word clouds to express the content of the data. Figure 5.3 shows a screenshot of QDA where the text is in the middle, the codes we created are on the left, and the emerging structure is on the right. We simply highlight a word or a phrase, and double click on the code that best represents it. There may be overlaps of meaning whereby one phrase may belong to two or more codes.

Figure 5.4 shows a frequency distribution of the codes generated in QDA Miner Lite. It shows that the text is dominated with examples of what promotes learning, followed by examples of what hinders learning, and then the other four codes.

(5) Reporting Results

Qualitative results should generally be presented in a way that captures the richness of the findings. Qualitative analysis is often aimed at deriving meaning from the data. Using direct quotes from participants is an effective way to powerfully convey meaning. Using software opens up the possibility to show tables and graphs that assist in understanding the meaning. It is, frankly, a delicate balance between capturing detail and summarizing findings. A bit of guidance can be found in Strunk and White's classic *Elements of Style*:

> Vigorous writing is concise. A sentence should contain no unnecessary words, a paragraph no unnecessary sentences, for the same reason that a drawing should have no unnecessary lines and a machine no unnecessary parts. This requires not that the writer make all his sentences short, or that he avoid all detail and treat his subjects only in outline, but that every word tell (Strunk and White, 1999, p. 18).

5.7 Summary

Quantitative data is expressed in numbers or quantity units, while other data is referred to as qualitative. In practice, one will often conduct a qualitative study with an exploratory design before proceeding with a descriptive design in

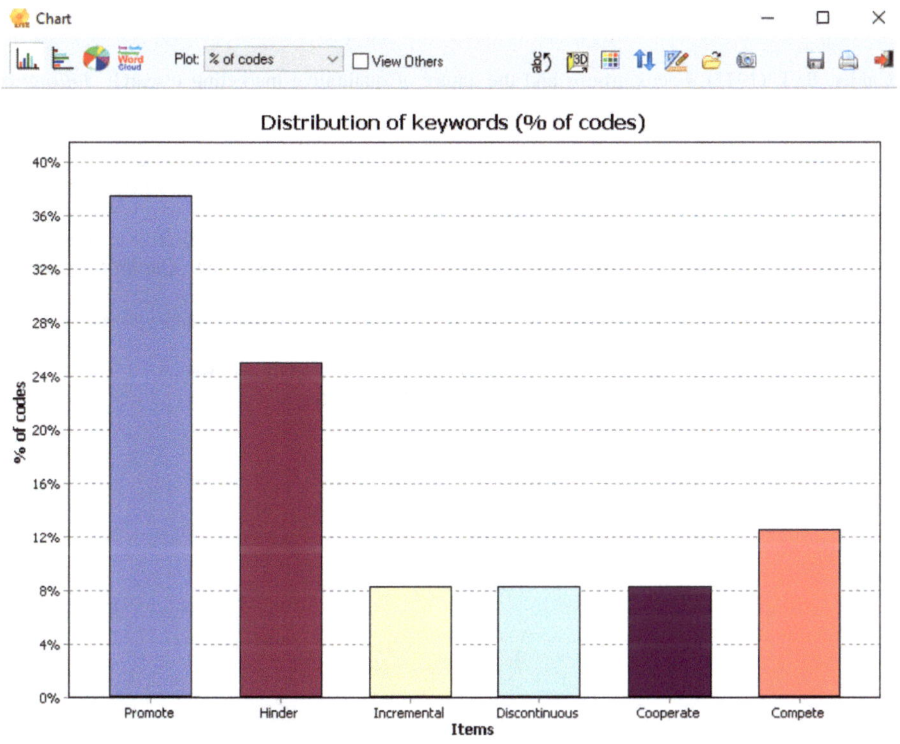

Fig. 5.4 Qualitative data analysis output

the form of, for example, a questionnaire survey. In some cases, one will settle for the qualitative study.

Although it is easy to imagine that social media is only about big data, social media has had an extraordinary rise in qualitative studies. A whole range of techniques and solutions are being developed to promote qualitative knowledge acquisition from the Internet and mobile phones.

Within focus groups, we distinguish between an exploratory approach, a phenomenological approach, and a clinical approach. Individual in-depth interviews are appropriate when the individual's personal experiences are of interest or the topic is sensitive. Projective techniques are based on methods used in clinical psychology. The purpose is to get the respondents to express their beliefs and attitudes in situations where they cannot or will not express them to direct questions.

Developing codes and applying a coding strategy facilitates qualitative data analysis. Software, like QDA Lite, greatly facilitates and enhances the coding process and presentation of the data.

References

Calder, B. J. (1977). Focus groups and the nature of qualitative marketing research. *Journal of Marketing Research, 14*(3), 353–364.

Gioia, D. A., Corely, K. G., Hamilton, A. L. (2013). Seeking Qualitative Rigor in Inductive Research: Notes on the Gioia Methodology. *Organizational Research Methods, 16*(1), 15–31.

Haire, M. (1950). Projective techniques in marketing research. *Journal of Marketing, 14*(5), 649–656.

Hofstede, A., van Hoof, J. J., Walenberg, N., & de Jong, M. D. T. (2007). Projective techniques for brand image research. Two personification-based methods explored. *Qualitative Market Research, 10*(3), 300–309.

Strauss, A., & Corbin, J. (1990). *Basics of qualitative research: Grounded theory procedures and techniques*. Sage.

Strunk, W. & White, E. B. (1999) *The Elements of Style, 4 ed.* Allyn and Bacon.

Yin, R. K. (2009). *Case study research: Design and method.* Sage.

Questionnaire Surveys 6

Contents

6.1 Introduction

As authors, our goal with this book is to provide the reader with research tools that improve decision skills. One way to measure how well the book is working could be to measure how engaged the reader is in the subject. We could ask the following questions (Fig. 6.1):

If it turns out that readers score high on all questions, except for question 3, we can directly work on techniques that help students to set ambitious yet realistic goals.

J. E. Sallis et al., *Research Methods and Data Analysis for Business Decisions*, Classroom Companion: Business, https://doi.org/10.1007/978-3-030-84421-9_6

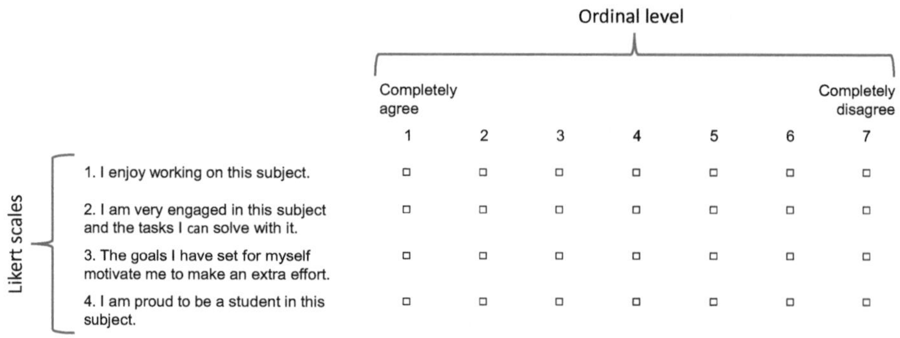

Fig. 6.1 Measurement scales for student engagement

In this chapter, we focus on how *questionnaire surveys* can be used to measure the variables that are derived from the research question(s). First, we need to clarify some terminology because sometimes these terms are used interchangeably. However, there are some distinct differences to keep track of. A *concept* is an abstraction of some sort of idea or phenomenon. This chapter began with discussing the concept of engagement. A *construct* is very similar to a concept. However, in social sciences a construct is more specifically defined and often related to measurement. Later in this chapter, we will talk about construct validity in terms of how well a construct is measured. A *variable* is something that can take on different values. With our example of the engagement concept, we could specifically define engagement as a measurable construct with different traits. The traits are represented by the four questions, which we can call measures or variables. To make things just a little more complex, in social sciences many constructs cannot be directly observed, in which case we call them *latent constructs*.

The questions in our engagement example are *operationalizations* of the latent construct, engagement. Sometimes, several questions are needed to measure a construct, while at other times we can settle for a single question. Our goal is to, through the use of the questionnaire, obtain reliable and valid measures of the constructs and variables we are interested in. This applies regardless of whether the data is collected on physical questionnaires, online, or through a mobile device.

In this chapter, first we look at how to operationalize and measure constructs and then validity and reliability in connection with measurements and the most important sources of error in questionnaire surveys. We discuss how to create a questionnaire, where we specifically consider question formulation and different scales. Finally, we discuss the pros and cons of the four most common ways to conduct questionnaire surveys (in-person, online, by telephone, and postal surveys).

6.2 Constructs and Operationalization

Most people who make a questionnaire are surprised to discover that it is far more demanding than they originally thought. The best way to learn is probably through experience. There are, however, some rules of thumb and general insights that can make the process easier. The steps that should be followed to measure concepts are shown in Fig. 6.2.

The first step is to make a *theoretical definition* of the constructs that will be included in the questionnaire. When writing a research paper, the theoretical definitions are usually in the introduction or theory section. We strongly recommend looking at relevant literature to see how the construct has been defined within the relevant scientific domain. Building on the engagement example, the definition could come from a dictionary. However, engagement within a specific research context is likely to be more nuanced than from a dictionary. This is important because the theoretical definition determines what is included within the construct, or in other words, the dimensions of the construct.

In the early 2000s, the Norwegian company Sweet Protection (sweetprotection.com) won an innovation award and two design awards in the extreme sports segment. Today, more than a decade later, let us say that the current management is interested to measure brand awareness across age, gender, and nationality. The research question is: "For Sweet Protection, how does brand awareness vary across age, gender, and nationality?"

Using a questionnaire, we want to measure:

- Brand awareness.
- Age.
- Gender.
- Nationality.

Age, gender, and nationality are quite straightforward to measure with the questions:

Fig. 6.2 Measuring constructs

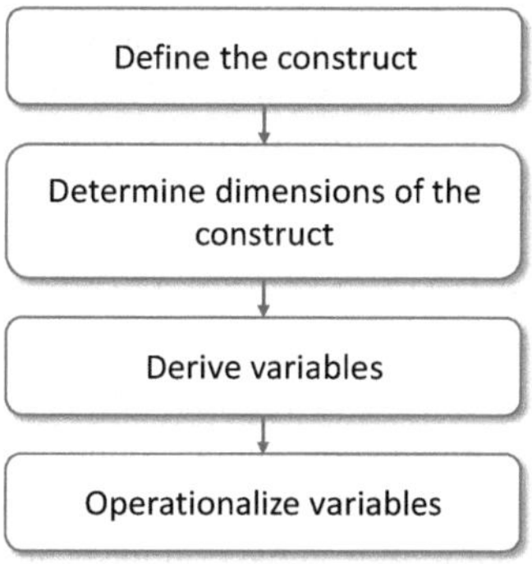

- How old are you?
- What gender are you?
- What nationality are you?

When it comes to the construct, brand awareness, it is more challenging. First of all, brand awareness is a latent construct. We cannot directly observe it, so we need to be creative in how we measure it. What do we mean by the brand awareness of Sweet Protection? If we directly ask whether the respondents have heard of Sweet Protection, we are *priming* their answers by providing the name. We can also call this *aided knowledge*. We can expect that many will answer yes, especially if they associate the product type with something they think they should know about. Any type of priming can greatly increase measurement error.

A more neutral approach would be to ask the respondents to first list all the brands they know within the relevant product category. This is what marketers refer to as the *evoked set* of brands and is an example of *unaided knowledge*. Then, ask them if they have heard of Sweet Protection. In practice, it is common to combine the two methods by first asking for unaided knowledge, and then aided knowledge. It is likely that the way and the order in which we ask will result in very different answers.

Researchers must justify all their choices in research methodology, including how constructs are defined and operationalized. Often, too little time is spent on defining constructs that will be operationalized in a questionnaire. Time pressure leads to choosing simple solutions. Focus groups or in-depth interviews can provide useful insights when it comes to mapping what content people put into different concepts. At a minimum, it is important to discuss operationalizations of constructs in order to get different perspectives than those you have yourself.

Operationalization is the process of translating theoretical constructs into something that is empirically measurable. Following the logic of the book, we have a research purpose that is refined into a research question(s). From the research question, we derive theoretical concepts that we define as measurable constructs. In order to answer the research question, we need to decide how to measure the theoretical constructs. We do this through operationalizing the theoretical constructs.

Let us take the example of the Norwegian graphics company, Vizrt (vizrt.com). In 2018, they had revenues of about 120 million euros generated across three global regions. If Vizrt wants to know how their business has evolved across regions in recent years, they could ask, "How has Vizrt regional revenue changed in recent years?" This requires that we operationalize Vizrt, regional revenue, and recent years.

At first glance, it may seem unproblematic to operationalize these theoretical constructs. After all, Vizrt is Norwegian and their financial statements show regional revenue for several years. However, what is regional revenue? This requires a definition of regions and a definition of revenue. Vizrt is based in Norway, so we might assume that revenues are defined according to Norwegian standard accounting practices. However, what if revenues are defined differently in other regions? For that matter, what countries are included in each region? Presumably, Vizrt may want

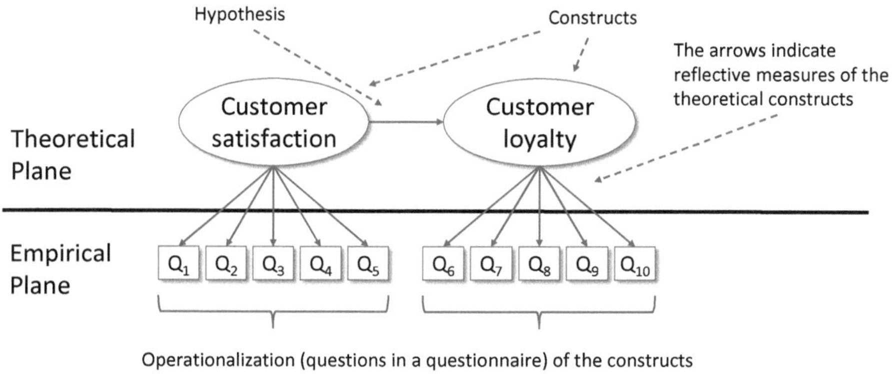

Fig. 6.3 From theory to empirical data

to benchmark its revenue development against other similar firms. To do this, they need to have comparable definitions of revenues and regions. Defining Vizrt can even be difficult since as they have globalized, they may have licensing agreements, import agents, and joint ventures that make it difficult to determine the boundaries of what is and what is not Vizrt.

The example shows that even relatively simple terms require careful consideration when determining how to measure a construct. *Measurement* is the process of recording observations according to rules. Many theoretical constructs are latent, and therefore, not directly observable. Even the relationships between the constructs may not be directly observable. As an example, we can draw a connection between "customer satisfaction" and "customer loyalty." The theory is that, all else being equal, the more satisfied a customer is, the more loyal he or she will be. If we want to test this theory, we need to define and operationalize customer satisfaction and customer loyalty into variables that can be empirically measured.

Bagozzi (1994) emphasizes that it is wrong to think of measurement only as operations to record a phenomenon. "Measurement achieves meaning in relation to particular theoretical concepts embedded in a network of concepts, where the entire network is used. This means that the measurement of the individual variables should not take place in isolation. Measurement operations for terms that belong theoretically should be viewed in context." Figure 6.3 illustrates the hypothesized relationship between the theoretical latent constructs *customer satisfaction* and *customer loyalty*. Moving from the theoretical plane to the empirical plane, the 10 questions in a questionnaire are the operationalizations of the latent constructs. This is a way to measure the level of customer satisfaction and the level of customer loyalty. If the measures are valid and reliable, meaning they accurately reflect the "true" theoretical concepts, then we can draw conclusions about the hypothesized relationship between the concepts.

It is common to say that reliability is a necessary but not sufficient condition for validity. In order to measure what we intend to measure, the measurements must also be valid. We aspire to have low systematic and random error. An example of systematic error is to forget an important question when measuring a concept. This means that the dimensionality of the concept is not fully represented, which systematically lowers the validity of the measure. Random error may be that some respondents, but not all, misunderstand the wording of one of the questions. It can be expressed in the form of a simple equation: $Xo = X_T + Xs + X_R$.

Xo: observed value.
X_T: true value.
Xs: systematic error.
X_R: random error.

The equation simply states that the observed value is equal to the true value plus systematic and random error. The highest possible validity is when the observed value equals the true value ($Xo = X_T$). If the measurement is perfectly reliable, then random error equals zero ($X_R = 0$). In practice, measurements are never perfect. Therefore, it is a question of degrees of validity and reliability in any research setting.

In social sciences, we are often measuring constructs such as attitudes, perceptions, and values. It is seldom possible to properly measure such complex constructs with a single question or observation. As an illustration, I often ask my students whether they think it is valid for me to measure their intelligence with a single question. Intelligence, just like many social constructs, is multidimensional and requires multiple measures. In Fig. 6.3, each of the two theoretical constructs is measured by five questions on the empirical plane. This is indicated by the arrows from the theoretical constructs to each question. The direction of the arrows pointing from the constructs to the questions means that the questions are reflections of the construct they measure.

6.3 Validity

Evaluating validity is difficult, and it is common to talk about many different forms of validity. Earlier in the book, at a general level we discussed the distinction between internal and external validity as well as reliability. In this section, we will look at different types of validity and reliability at the measurement level. Specifically, we focus on the validity and reliability of the measurement instruments we use, with focus on questions in a questionnaire. We will talk about content validity, construct validity (convergent and discriminant), face validity, and statistical conclusion validity. We end with reliability.

6.3.1 Content Validity

Content validity is the extent to which the measurement captures the entire theoretical construct. For example, if we want to measure interpersonal trust between two coworkers, first we need to look at the relevant literature and see how interpersonal workplace trust is defined. There will be at least two dimensions. One will be the calculative dimension where the two coworkers trust (or do not trust) each other based on an objective evaluation of professionalism. Another will be the emotional (affective) trust that builds up through contact and friendship. For the sake of content validity, to capture the content of interpersonal workplace trust, there will need to be at least two measures: one to measure calculative trust and the other to measure affective (emotional) trust. Many constructs have validated measures in published empirical articles. It is best to use established measures and if necessary adapt them to your context. It is not a good idea to invent new measures unless there is nothing available. If there are no established measures, use conceptual definitions, focus groups, or interviews to determine the content of a construct for your population and then create the measures.

6.3.2 Construct Validity

Construct validity is of particular importance when measuring several constructs and the relationships between them. Carmines and Zeller say, "Fundamentally, construct validity is concerned with the extent to which a particular measure relates to other measures consistent with theoretically derived hypotheses concerning the concepts (or constructs) that are being measured (1979, p. 23)." It is particularly relevant when they are latent constructs. *Construct validity* is the degree to which measures are related to a specific construct and not related to other constructs. It is expressed and evaluated as two subdimensions:

1. *Convergent validity* is the extent to which specific measures converge on a construct. Measures of the same construct are highly correlated.
2. *Discriminant validity* is the extent to which the same measures do not converge on other constructs. Measures of one construct should not highly correlate with measures of another construct.

In the example in Fig. 6.3, construct validity will be achieved if questions 1–5 converge on customer satisfaction and not on customer loyalty (they discriminate), and questions 6–10 converge on customer loyalty and not on customer satisfaction (they discriminate). We discuss and show how to measure construct validity in Chap. 12 on factor analysis.

6.3.3 Face Validity

Face validity is the simplest measurement assessment and should not be confused with content validity. *Face validity* is the subjective assessment of whether the measures are measuring what they are supposed to measure. Do the measures make logical sense? Are the measures so obvious that people generally agree on them? A good way to establish face validity is to ask experts or people who are familiar with the concepts. You could ask a small group of respondents to comment on your questionnaire before you do the full survey.

6.3.4 Statistical Conclusion Validity

Statistical conclusion validity is the extent to which conclusions are supported by statistical analysis, and is often connected to conclusions about covariance. If we have high inconsistency in the measurement instrument, in the setting, and in data processing, we have low reliability which weakens statistical conclusion validity. As we will discuss in Chap. 9, the significance level set for a study will affect the danger of making Type II errors. In other words, the test does not have enough strength to be valid with regard to statistical conclusions. The appendix to Chap. 10 shows a number of assumptions that must be met in order for us to draw conclusions from statistical findings.

6.4 Reliability

Reliability is the extent to which a measurement produces consistent results when repeated. All measurements are subject to random error. A measurement with low random error has high reliability. If we measure customer satisfaction with a questionnaire, there will be random factors that influence how respondents answer. If we immediately measure it again with the same respondents, assuming that nothing substantive has happened to change it, the results will be slightly different. In general, if respondents have well-developed opinions about what we are measuring, repeated measures will be quite similar. If they do not have developed opinions, they effectively guess the answers, and randomness increases. Reliability, in this sense, is not just how good the measurement instrument is (the questionnaire), but also a function of the context and respondents. Later in the book, we will show how to calculate reliability through a measure of the internal consistency of multiple measures, Cronbach's alpha. Together with factor analysis, we can assess the quality of a questionnaire.

The four types of validity (content, construct, face, and statistical conclusion) and reliability have all been discussed in terms of assessing a measurement instrument, which in this case is a questionnaire. The concepts of validity and reliability can, however, also be applied to the entire research design. As far as experiments are concerned, we mentioned earlier that internal and external validity are of primary importance. We discussed the advantages and disadvantages of different research designs (e.g., laboratory or field experiments). If we refer back to the equation we used to express validity ($X_O = X_T + X_S + X_R$), it is crucial to, as much as possible, avoid random and systematic error. Sample type and sample size are important for evaluating error in surveys. With a *probability sample* it is possible to measure random error, which is not the case for a *non-probability sample*.

Referring back to our discussion in Chap. 1 about the philosophy of science, a staunch constructivist would reject what we have just written about validity and reliability. We have been implicitly taking a positivist perspective by suggesting that validity and reliability exist independently of the phenomenon and that we can, to some extent, objectively measure them. This is a very good and dare I say valid point to keep in mind. We are studying social phenomena in highly complex systems. People, what we have been calling respondents, subjectively experience the world. How can we measure and draw conclusions, which we usually generalize, about people without accounting for their personal context and experience? How valid is the positivist perspective?

Herein lies the (healthy) tension between positivism and constructivism. Do not be fooled by numbers (none of the polls showed Trump being elected). Yet at the same time, there are consistent beliefs and behaviors at the group level. Constructivism and positivism work hand in hand through exploratory, descriptive, and causal research designs in a perpetual fluctuation between exploring and testing ideas.

6.5 Measurement Scales

Measurement entails recording the amount of something, which could be characteristics of a person or a business. The characteristics are *variables* and, depending on their nature, they can be measured on four types of scales (nominal, ordinal, interval, and ratio). This is called the *level of measurement*. At the lowest level (nominal), we measure the frequency of discreet variables. There is no ranking; they are either present or absent. At the highest level (ratio), all arithmetic functions are possible so we can do things like calculate averages. A good rule of thumb is to always measure at the highest level possible. We will go through examples of Likert and semantic differential scales because they are commonly used in questionnaires (Fig. 6.4).

Nominal Level

Nominal level scales only provide the basis for grouping observations into different categories. Classic examples for individuals include gender and place of residence;

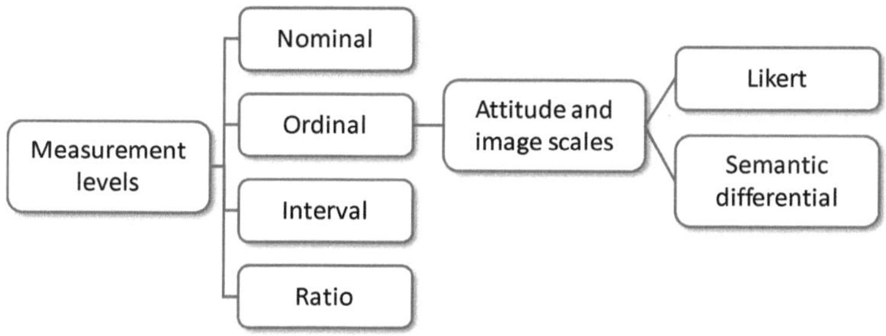

Fig. 6.4 Measurement levels

Fig. 6.5 Nominal scale example

Where is the head office of your company?

☐ Denmark

☐ Finland

☐ Norway

☐ Sweden

☐ Other

for companies, industry affiliation and localization. The categories in these variables are states of being that have no ranking of one being above or below the other. Frequency is meaningful in as far as there could be, for example, more head offices in one country than another (see Fig. 6.5).

We suggest to always measure at the highest level possible. Through a bit of creativity, it is sometimes possible to move up. For example, from a list of ten personality characteristics, you could ask respondents to choose the four most important characteristics for a good manager. Strictly speaking, this is a nominal measure because, at best, you can only show the frequency of the choices. There is no ex ante ranking of characteristics and some frequencies may be equal. You can argue that it is an ordered frequency. However, it is still an unsophisticated measure. Instead, you could list the ten personality traits and for each trait ask the question, on a 1–7 scale (1 is low, 7 is high), how important is this characteristic for being a good manager?

This results in an ordinal measure, which we explain shortly can be treated as *approximately interval.*

Ordinal Level

Ordinal level scales have categories that can be ranked in a specific order. They do not necessarily say how much one category is relative to another category. In questionnaires, it is common to ask a person's level of education and suggest categories like (1) elementary school, (2) high school, and (3) university education. The categories are in a clear order; however, the number of years within each category may vary greatly across education programs and regions. It is wrong to try and calculate the average education from this kind of scale.

Another example of an ordinal scale is when respondents are asked how much they agree or disagree with some statement (see Fig. 6.6).

Strictly speaking, these are ordinal scales. Each category from 1 to 7 shows the ranking, not the distance within or between each category. Respondents may feel somewhat nonchalant about the difference in choosing 3, 4, or 5. However, going from 5 to 6 may seem like a substantial choice, and the distance between 6 and 7 even more extreme. Nevertheless, these sorts of ordinal scales are often treated as though they are continuous variables (approximately interval).

	Completely disagree 1	2	3	4	5	6	Completely agree 7
I intend to buy this product	□	□	□	□	□	□	□

Fig. 6.6 Ordinal scale example

How old are you?	________
How much do you earn per month?	________
How many times have you been in the store this month?	________

Fig. 6.7 Ratio scale example

Interval Level

Interval level scales are ranked and the value between each level is meaningful. If we knew the distance between each number in the Fig. 6.6 ordinal scale example and the distances were equal, then it would be an interval level scale. There is no need for an absolute zero point on an interval scale.

The classic example of interval scales is temperature. In the USA, temperature is traditionally measured in Fahrenheit, while most other countries in the world use Celsius. The scale developed by Anders Celsius at Uppsala University has zero as the freezing point of water and 100 as the boiling point of water. Daniel Fahrenheit's scale has 32 as the freezing point and 212 as the boiling point of water. Fahrenheit defined zero as an equal mixture of water, ice, and salt. Today, a healthy adult human should have a temperature of 98.6 Fahrenheit or 37 Celsius. We can calculate the average temperature for a series of observations regardless of which scale we use, since the distance between each value is equal within each scale. However, so long as there is no natural zero point, we cannot say that, for example, 20 °C is "twice as hot" as 10 °C. We can convert a temperature given in Celsius to Fahrenheit by multiplying the Celsius temperature by 1.8 and then adding 32: $(10 \times 1.8) + 32 = 50$ F and $(20 \times 1.8) + 32 = 68$ F. 68 is not twice as hot as 50!

Ratio Level

Ratio level scales have the same characteristics of an interval scale plus an absolute zero. This means that it is possible to say how many times greater one value is compared to another. Variables such as age, income, and the number of store visits all have an absolute zero point. It is correct to say that someone who is 40 years old is twice as old as someone who is 20 years old. A person who earns 50 thousand euros per year earns 5 times a person who earns 10 thousand euros per year. A person who has been in a store six times this month and three times last month has double the number of visits this month (Fig. 6.7).

6.5.1 Parametric Versus Nonparametric Methods

When choosing how to measure something, it is important to consider what methods will be used to analyze the data. *Parametric statistical methods*, which encompass most of the more advanced statistical methods, are contingent upon measures at the interval or ratio level. Strictly speaking, parametric statistical methods should not be applied to ordinal or nominal level data. This is why we suggest the rule of thumb to always measure at the highest level possible. Interval and ratio level data can always be transformed into ordinal or nominal groups. However, it is impossible to convert a lower level measure to a higher level.

One caveat to the rules around measurement level are scales called *approximately interval*. We referred to this in the earlier section on ordinal scales. Often, the Likert-type scales, like the one shown in Fig. 6.6, are treated as though they are at a higher level of measurement; they are approximately interval. This is so that the variables can be used with the wider array of parametric statistical methods.

6.6 Attitude and Perception Measurement

Attitudes are formed over time as a result of how a person perceives phenomena. They are not directly observable and therefore challenging to measure. Nevertheless, research often requires measuring attitudes and perceptions. Single questions do not capture their dimensionality, so most often, a battery of questions is used to measure them (see Fig. 6.8). Each question is meant to measure a slightly different aspect of the construct. Generally speaking, this is the case for all complex social constructs.

There are many types of question batteries. Two types of questions have become dominant in social sciences: *Likert scales* and *semantic differential scales*. We will give a brief description of these two scales with special emphasis on measurement level.

Likert Scales

The Likert scale is named after the psychologist Rensis Likert (1932), who originally described this way of measuring attitudes. Respondents are asked to indicate to what extent they agree or disagree with a series of statements about the phenomenon one wants to measure. Historically, a response scale with five response options is used. Researchers commonly use more response options (6, 7, and 10), and often only use

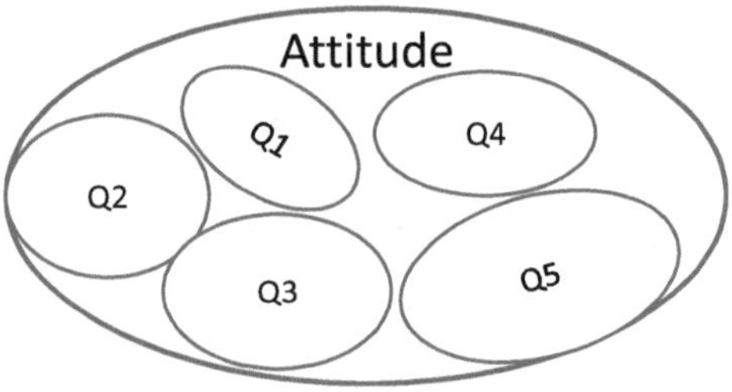

Fig. 6.8 Using several questions to measure a construct

Fig. 6.9 Classic Likert scale

Strongly agree	Agree	Neither agree or disagree	Disagree	Strongly disagree
☐	☐	☐	☐	☐

anchors like strongly agree and strongly disagree, leaving the middle options blank (Fig. 6.9).

Scale development proceeds as follows:

1. Write theoretical definitions of each construct.
2. From existing scales in the literature and qualitative research, compile a list of questions that fully cover the dimensions of the constructs.
3. Edit the questions based on advice from available experts.
4. Test the questions on a small sample of respondents and ask them to comment on the scales.
5. Run simple statistical analyses. Correlation (Chap. 8) and reliability (Chap. 8), and if the sample size permits, factor analysis (Chap. 12).
6. Based on the results, edit the questions and finalize the questionnaire, or if necessary, repeat the process.

The list of questions should always be pre-tested on a small sample prior to the main survey. Look for questions where all the respondents answer in the same way, and revise or remove them. When there is no variation in the answers to a question, the question is not capturing any *variance* in the perceptions you are trying to measure. You should also test the *covariance* between questions through *correlation analysis*. When questions are supposed to reflect the same construct, they should be highly correlated. Questions for different constructs should have low correlations. The set of questions for each construct should be tested for reliability with Cronbach's alpha. The generally accepted level of reliability for the measures of a construct is 0.7 (Nunnally, 1978). Questions that do not correlate as they should or that are not reliable should, under most circumstances, be removed from the questionnaire. The remaining questions are included in the questionnaire to be used in the main survey. Keep in mind that after editing, the dimensions of the theoretical constructs must be sufficiently represented. If not, the process needs to continue or the definition adjusted.

In the final version, some researchers choose to mix positive and negatively worded questions to stimulate the respondent's engagement. For example, with the anchors strongly agree and strongly disagree, a respondent could be asked in a positive way, "I trust this company," or in a negative way, "I do not trust this company." If a construct is measured with both positive and negative wording, then one of the scales must be reversed before analysis. If you do not reverse either the negative or the positive, then they cancel each other out when analyzing them together.

Mixing in reversed questions presents a dilemma, especially for long questionnaires. Respondents have limited time and energy for answering surveys. If the survey is sent to managers on the job, they may be stressed to complete it quickly. Respondents may not cognitively process that a few random questions are negatively worded, which will cause them to answer either completely opposite to what they mean, or they may just get confused and either leave the question blank or guess at a random answer. If you choose to include negatively worded questions, be particularly aware that they may have much higher variance or missing values when compared to the positively worded questions in the same scale. If this is the case, they should be removed from the analysis.

Alternative scale values can be used, for example, −2, −1, 0, 1, 2, and more response options like 7-point or 10-point scales. Generally speaking, look at how the construct has been previously measured and which scale values they used. Then, determine which values best suit your purpose. Using scales with no mid-point, like a 6-point or 10-point scale, forces the respondent to take a position toward agreeing or disagreeing. The advantages of this must be weighed against the disadvantages of respondents not being able to be ambivalent about a question if that is what they would choose to do. Being forced into taking a position can frustrate respondents and increase non-responses.

We have already mentioned that many researchers treat these types of ordinal scales as though they are metric. For statistical analysis, this is problematic because of the possible unequal distances between response points and that many statistics assume normally distributed data. If you choose to compromise on how to treat ordinal data, then it will ease the problems by having more response options, like seven or ten, and measuring a construct with multiple questions. When multiple questions are reliable and valid, it is common to add the questions in each construct together to create an aggregated measure. Aggregated measures with more response options tend to have less error and are more normally distributed than individual measures. Nevertheless, you are only alleviating the symptoms and you must always remember that you are *breaking the rules* by treating an ordinal scale as though it is metric!

Semantic Differential Scales

The second scale, which has become a classic in business research, is called the semantic differential scale. The scale was originally used to measure the underlying structure of words and the meaning of words (Osgood et al., 1957). It was later adapted to measure *image*, which in business terms pertains to *brand image*. The measurement process starts with determining the *attributes* that make up the image. For each attribute, you assign two polar anchors for the scale. Using the example of price, you can either have *monopolar* (high price–low price) or *bipolar* (expensive–cheap). Typically, respondents are asked to decide how they perceive the object by placing it on a scale between the two extreme values for each attribute. Often, a 7-point scale is used where only the two anchor values are named (see Fig. 6.10). It is not uncommon to use 20–30 attributes in a survey; however, the more attributes, the more complex it becomes.

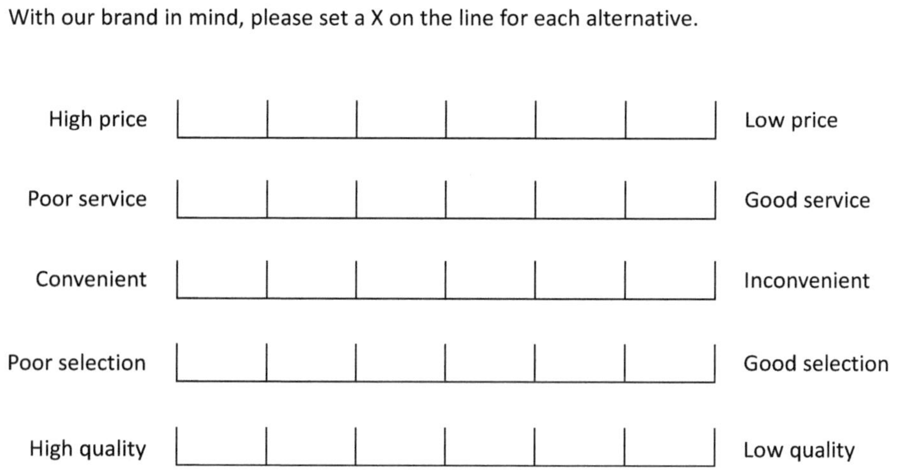

Fig. 6.10 Semantic differential scales

Semantic differential scales, like Likert scales, are ordinal level measures. They face the same challenges when being treated as though they are interval level. Instead of using average values, what we call the *mean value* in statistical terminology, some researchers recommend using median values. The *median* is the middle value when observations are arranged in either an ascending or descending order. Compared to the mean, the median is not sensitive to deviations from normality and extreme values. In practice, however, it is common to assume that the intervals are equal and calculate the mean (the arithmetic average) based on the values in the scale. Unlike Likert scales, where several questions are added together to form an aggregated construct, results for semantic differential scales are often displayed attribute by attribute with a snake diagram connecting the center point for each attribute (see Fig. 6.11). They are often used to compare two competing brand images on attributes that are important for target groups. In the example in Fig. 6.11, the high perceived price (Our brand) is compared to the lower perceived price (Competitor brand). It could be troubling that despite the perceived price differential, they are perceived about equal on service. All the dimensions could be addressed by different managerial actions.

6.7 Scale Values

Questionnaires contain many types of questions about facts (e.g., demographic questions like age and gender), behaviors (e.g., Internet versus brick-and-mortar shopping habits), knowledge (e.g., brand awareness), and attitudes, opinions, and perceptions. The challenge for any researcher is to develop a questionnaire that will get valid and reliable results. In this section, we discuss different types of scale values.

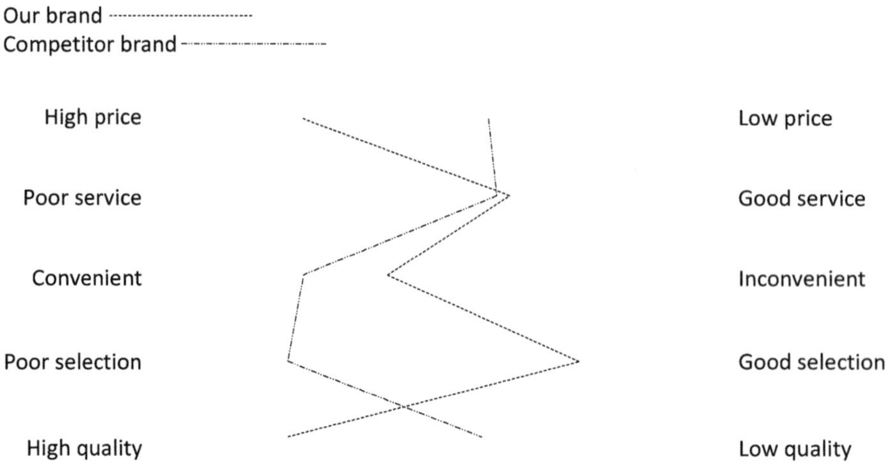

Fig. 6.11 Semantic differential results example

Fig. 6.12 Comparative and non-comparative scales

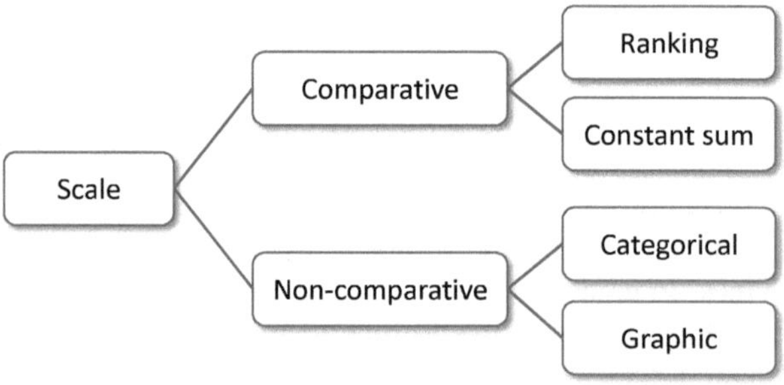

It is common to distinguish between single-item and multiple-item (composite) scales. The composite scales are question batteries that capture different dimensions of a construct as it is theoretically defined. As we have already discussed, Likert scales and semantic differential scales are extensively used to measure single and multiple dimensions. Keep in mind that a composite scale could be broken down and defined as a set of single-item scales. Dimensionality is a question of definition. For example, learning is often defined with three dimensions of (1) information sharing, (2) information processing, and (3) memorization. Alternatively, we could treat each of the dimensions of learning as an independent construct.

Two distinctive types of scales are comparative and non-comparative. *Comparative scales* ask respondents to compare two or more alternatives (like two products) with each other when responding to the questions. *Non-comparative scales* ask respondents about a single phenomenon (like a product). Within each type there are many variations, of which we mention some of the most important (Fig. 6.12).

If you were to invest a relatively large amount of money, which communication channel would you prefer?

Please rank the following alternatives from 1 to 4, where 1 is the channel you prefer most.

A. Face-to-face with a financial advisor. ☐

B. Telephone contact with a financial advisor. ☐

C. Electronic communication by email/internet. ☐

D. Traditional mail communication. ☐

Fig. 6.13 Ranking scales

Comparative Scales

Comparative scales ask the respondent to compare alternatives, either within the scale, or like our semantic scale example, comparing two alternatives against each other on different attributes. The *ranking scale* is the most common example of a comparative scale. Respondents are asked to rank set of options based on given criteria. An example may be that a bank wants to know which communication channels are most important for a customer who wants to invest a relatively large amount of money (see Fig. 6.13).

The ranking scale generates ordinal level data. We cannot say how much better one alternative is relative to another, only that one alternative is preferred over the other. If the number of alternatives is large, respondents have difficulty ranking them. They may care about a few alternatives, but not all of them. When there are several alternatives, one solution is to ask the respondent to pick the top three or four, and rank them. Even when some respondents are ambivalent, the scale forces them to prioritize, which could negatively influence the validity of the ranking. In some contexts, however, the ranking scale provides useful information.

A refinement of ranking is the *constant sum scale*. Respondents are told to allocate a fixed number of points, such as 100, between a set of alternatives. Not only does this provide the ordinal ranking, but also information about the difference between the alternatives. Continuing the investment example in Fig. 6.13, the respondent is instead asked to allocate points (see Fig. 6.14). In the example, we show what a respondent has answered on the scale. With 75 points in the face-to-face alternative, the respondent is clearly showing his or her overwhelming preference for meeting in person to discuss large investments. Similar to the simple ranking scale, if there are too many alternatives the respondent may have difficulty with the allocation. Keep in mind that these kinds of scales are difficult to administer during interviews because the respondent may want time to consider the alternatives, and the presence of the interviewer (in person or on the telephone) may stress the respondent and negatively influence the validity of the answers. Finally, although the scale is numerical, capturing information about the distance between alternatives, it is debatable as to whether it can be treated as an interval level scale.

If you were to invest a relatively large amount of money, which communication channel would you prefer?

Please allocate 100 points between the alternatives, where more points indicates a greater preference.

A. Face-to-face with a financial advisor.	75
B. Telephone contact with a financial advisor.	5
C. Electronic communication by email/internet.	5
D. Traditional mail communication.	15
	Sum 100

Fig. 6.14 Completed constant sum scale

	Completely agree 1	2	3	4	5	Completely disagree 6
My boss fully supports my decisions.	□	□	□	□	□	□

Fig. 6.15 Non-comparative Likert scale

Please mark an X on the line which indicates your level of agreement.

	Completely agree	Completely disagree
My boss fully supports my decisions.	├───────	───────┤

Fig. 6.16 Continuous rating scale

Non-comparative Scales

Non-comparative scales evaluate things in isolation, not explicitly compared to anything else. The *Likert scales* we talked about earlier are often non-comparative. The respondent is given a limited number of response alternatives, which he or she is to choose from (see Fig. 6.15).

In this example we have only labels the anchors, and by not having a neutral middle value we are forcing the respondent to take a position in either direction. If there is a risk that the question does not apply to all respondents, you can also provide an answer alternative with "does not apply" or "do not know." The scale is balanced, in that there are an equal number of positive (agree) and negative (disagree) answer alternatives.

Continuous rating scales provide a line upon which the respondent is asked to put a mark. The point of this scale type is that respondents indicate their perceptions as a continuous variable on a line. The researcher then applies numerical values based on distance from the origin (anchors). They are rarely used in practice because their reliability has been shown to be quite low (Fig. 6.16).

6.8 Question Formulation and Order

So far, all our examples have been of structured scales where respondents are given closed-answer options. Scales can be divided into:

(a) Open-ended responses, where the respondents answer in their own words.
(b) Closed-ended responses, where the answers are given as alternatives to choose from.

The great *advantage* of open-ended questions is that the respondent expresses their opinions in the way they want to. This sometimes encourages the respondent's engagement, so starting a questionnaire with open questions can be advantageous. It also helps the analysts to build a deeper understanding of each respondent. It is also good when there are too many alternatives to be expressed in closed-ended questions, or when the researcher does not have a good understanding of the answer alternatives. Closed-ended questions are based on the researcher's worldview, which may not reflect the respondent's worldview.

Despite all the advantages of open-ended questions, they are, relatively speaking, rarely used. This is because there are so many *disadvantages*. The responses will largely depend on how skilled the respondents are at expressing themselves, and how skilled the analyst is to interpret the answers. *Data coding*, which is the interpretation process done by the analyst, is immensely time demanding. This also leads to the dilemma of deciding how detailed to be when presenting the results. Some years ago, a large sample of Norwegian grocery store managers were asked, "What do you think is the most important problem facing grocery retailers today?" After coding, there were eight main categories and 39 subcategories. It was extremely time-consuming and the results were so diverse that its value was limited (Fig. 6.17).

With open-ended questions, the respondents will often emphasize different aspects of what they are asked about, which makes comparing answers difficult. Most of the time, an important aspect of surveys is to get comparable answers and, to some degree, generalize the results. Closed-ended questions make this vastly easier and faster, so they are the dominant form in questionnaire surveys.

Some questions are best asked as open questions. For example, if we want to know the age of a respondent it might be better to directly ask than to provide a set of

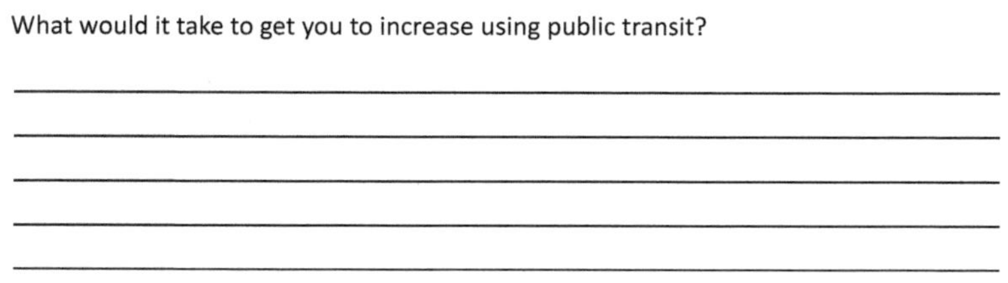
What would it take to get you to increase using public transit?

Fig. 6.17 Open-ended questions

Fig. 6.18 Ordinal question for ratio level variable

How many times have you used public transit during the past 30 days?

A. None ☐
B. 1-5 ☐
C. 6-10 ☐
D. 11 or more ☐

Fig. 6.19 Guidelines for making questions

1. Use simple and clear words.
2. Avoid leading questions.
3. Avoid implicit alternatives.
4. Avoid implicit assumptions.
5. Avoid generalizations.
6. Avoid double questions.

alternatives. This raises an important issue around measurement levels. A rule of thumb is to always measure at the highest level possible. Age is a ratio level variable. When posed as a set of timespans, like 20 and under, 21–40, 41–60, and 61 and over, it is reduced to being an ordinal level variable, thus restricting the analytical options. The problem can be alleviated by making each category the same number of years. However, unless you have a good reason, simply ask for age. A reason for creating categories could be the inability to provide an exact number. Many people could not answer exactly how many times they used public transit in a month. Instead, you may want to create meaningful categories that correspond to, for example, non-user, light user, medium user, and commuter. The categories could be related to definitions by transit authorities. When using specified response options, it is also important that the alternatives cover all possible answers. This can, of course, be achieved by including "Other" as a category. However, most of the responses should preferably not be in the "other" category (Fig. 6.18).

6.8.1 Question Design

Regarding the actual design of the questions, as opposed to the answer options, most textbooks offer guidelines. The most important are summarized in Fig. 6.19.

Use simple and clear words Which words used in a questionnaire will of course depend on who the respondents are. Nevertheless, a rule of thumb is to avoid unusual words and complicated concepts. In a survey among economists, one might ask whether they are for or against the European Central Bank following a fixed-exchange-rate policy. Clearly, this question would not be appropriate for the general population. If you did ask the general population, the answers will be based on different interpretations or simply guesses. This will increase the error variance in the data. Surveys show that many words and phrases that may seem simple to professionals are not understood by non-professionals. It is important to try to familiarize yourself with the respondent's way of thinking when formulating the questions. After all, the answer to the question starts with the respondent interpreting what is in question.

Examples of unclear words are "rare" and "often." Respondents may attach very different meanings to words like this. Problems may arise especially if such words are used on response alternatives. (Do you "often" or "rarely" go to the theater if you go 4–5 times a year?) Unclear words are often used in the question formulation itself. For example, during the 2017 European refugee crisis there were surveys with the question, "Would you temporarily allow a refugee to live with you?" The word "temporarily" in this context is rather unclear, and may be interpreted very differently. Even the word "refugee" is unclear. A young person or an old one, female or male? The lack of clarity can greatly influence the answer.

Avoid leading questions A leading question gives the respondent an indication of which answer he or she should choose. A national television broadcaster asked caregivers for the elderly, "To what extent might lack of resources be a problem for providing proper and dignified care?" They then reported, "Two of three caregivers say that the elderly in nursing homes receive unworthy treatment because of lack of resources." Sometimes, it is not so obviously leading. Asking, "Have you heard about product X?" will lead some respondents to answer yes, even when they have not. It is better to list alternatives and then ask which ones the respondent has heard of.

Avoid implicit alternatives An implicit alternative occurs when a possible alternative is not specified in the question. For example, "Are you satisfied with using public transit to get to work?" The only option is public transit, while there could be many alternative ways to get to work. Many studies have shown that the distribution of responses can be very different if one specifies the alternatives to be considered. If the purpose is to find out something about how the respondents consider public transit compared to the use of a private car, one should instead formulate the question: "Compared to using a private car, how satisfied are you with using public transit to get to work?" Have, for example, a 7-point scale anchored by very "dissatisfied" and "very satisfied," and an option for "does not apply" for respondents who do not make a choice between these options.

Avoid implicit assumptions An implicit assumption occurs when the consequences of the answer are not clear in the question. If you ask, "Are you for or against more resources for elderly care?" The question does not account for limited resources and trade-offs. For example, "Would you be willing to pay 2% more income tax to increase resources for elderly care?" This question specifies the consequence of being for or against increasing resources for elderly care.

Avoid generalizations By this we mean avoid broad statements that apply to a group. For example, "How can we make people less cynical toward politicians?"

This question makes the generalization that people are cynical toward politicians, which is not the case for everyone.

Avoid double questions A double question implies that the respondent is asked to relate to two phenomena within one question. An exporter who is asked, "Are language and culture challenging when doing business in Japan," might speak the language well, however, find the culture challenging. If the question is related to a scale of how challenging, then the exporter cannot properly express an answer for both phenomena.

Regarding question order, a rule of thumb is to start with simple and interesting questions to warm the respondent to the topic. Background questions about age, place of residence, income, and the like are usually included last. Some researchers use the *funnel method*. This means asking general questions first and then moving to questions that are more specific. *Priming* happens when early questions affect answers to later questions. For example, if respondents are first asked about specific political scandals, and later asked about their general attitudes toward politicians, they may be primed to be more cynical in their later responses. Priming, if used carefully, can also help activate memories around a topic.

We do not go into detail here; however, there are several things to consider for getting valid data and good response rates. The questionnaire's physical layout should be clear. Different web solutions have standardized formats that are well tested. You can possibly increase response rates by offering small gifts or a chance at a gift. A good introduction letter can engage the respondent. Be clear and concise, and do not waste the respondent's time with a long, detailed letter. Make sure it is easy for the respondent to submit their answers. When doing traditional mail survey, provide an addressed and postage paid envelope.

6.8.2 Pre-test

We cannot emphasize enough the importance of pre-testing the questionnaire with a small group before doing the main survey. Almost invariably you will discover small and sometimes large improvements that need to be made. Ideally, the pre-test respondents should be randomly chosen from the target audience, but at a minimum, the questionnaire should be tested on 5–10 people to discover if there are ambiguities and shortcomings that were overlooked.

6.9 Collecting the Data

We will discuss four ways of collecting data: (1) personal interviews, (2) online solutions, (3) telephone interviews and, (4) postal surveys. Fairly early in the process, it is important to decide how the data will be collected because it will determine how the questionnaire should be designed. For example, in telephone interviews, it is not reasonable to expect respondents to be able to evaluate more than three to four alternatives in one question. In personal interviews, the respondents may be shown pictures or other visual aids to facilitate the interview. With online or postal surveys, the respondent has time to evaluate several alternatives. The scope of the questionnaire, the number of questions, and their complexity also determine which way to communicate with the respondents. Each format has strengths and weaknesses.

The quality of the data from any of these methods is highly dependent on who the respondents are. We will discuss sampling in detail in Chap. 7. For the sake of explaining data collection, we clarify a few terms here. The *population* is defined by the researcher as the group that is being studied in the research. The *sampling frame* is the list of all possible respondents in the *population*. When the population is large, researchers often take a *sample* of a small part of the population. What they learn from the sample they try to infer is representative of the population.

6.9.1 Personal Interviews

Some *advantages* of personal interviews:

- Respondents may be shown relatively complicated visual stimuli.
- The interviewer can assist the respondent by explaining difficult questions.
- The interviews can be relatively long and comprehensive.
- The interviewer can persuade the respondent to complete and answer all questions.
- The interviewer can observe who responds and records reactions.

Some *disadvantages* of personal interviews:

- The interviewer's presence can influence how the respondent answers.
- The interview is time-consuming.
- Personal interviews take a lot of resources for each respondent (depending on duration).

It is difficult to generalize about time and costs. It depends very much on the way respondents are selected and the duration per respondent. Doing street intercepts for brief questionnaires is fairly effective for getting a lot of data quickly. On the other hand, interviewing managers who are in different locations, even different countries, results at best in one or two interviews per day.

6.9.2 Online Solutions

The *advantages* of online solutions:

- Inexpensive.
- Flexibility (the respondent answers when time and opportunity permit).
- Several free and paid solutions are available.
- The questionnaire can be tailored to how the respondent is answering (directed to different questions).
- Few physical boundaries, so long as respondents have open internet access.
- Online interviews and focus groups are possible.
- Many visual and audio aids are available.

The *disadvantages* of online solutions:

- The response rates may drop due to a virus hazard.
- It is easy to skip the survey.
- Bias in the sample.

Online solutions have a major advantage over the other solutions in cost and reach. No other solutions can reach so many people over such great distances so quickly at so little cost. This is why professional data collection agencies are switching to cloud-based online solutions. There are challenges with getting stuck in spam filters and respondents are reluctant to click on links due to the fear of catching a computer virus. Generally speaking, it is better to approach respondents who are connected to some sort of mailing list that gives the survey legitimacy. When randomly mass mailing people, there is little control over who actually answers so the results are likely quite biased. Without some legitimacy in approaching respondents, even if you get past the spam filter, it takes them a millisecond to push "delete" and you get no response.

Online solutions excel in providing audio and visual support to surveys. Focus groups and in-depth interviews can take place online with all participants in different locations. Panels can be managed online. There are data collection agencies that can put your questions online and provide constant real-time feedback on the responses. One particularly popular online survey tool is Survey Monkey. They offer free and pay services.

6.9.3 Telephone Interviews

Some *advantages* of telephone interviews:

- They can be conducted quickly.
- They are inexpensive compared to personal interviews.
- Easier to make contact than in-person interviews.
- The interviewer can clarify misunderstandings.
- The influence of interviews is less than in-person interviews.

Some *disadvantages* of telephone interviews:

- You cannot use complex scales.
- Visual stimuli cannot be used.
- The respondent does not have time to reflect on answers.
- The interview cannot be long because the respondent becomes impatient.

With the disappearance of telephone directories connected to fixed addresses, geographic sampling is increasingly difficult. Lists can be created from things like memberships in associations or computers can randomly generate telephone numbers in a geographic area. Yet, at the same time, more options to opt out of direct marketing are limiting the usability of these lists. Company registries are quite visible in various databases or from branch organizations. Companies may be easier to identify, although this does not mean they are willing to be interviewed.

6.9.4 Postal Surveys

Some *advantages* of online/postal surveys:

- Fairly inexpensive.
- Flexibility (the respondent answers when time and opportunity permit).
- Easier to pose sensitive questions.
- Relatively complicated question and response scales can be used.

Some *disadvantages* of online/postal surveys:

- Low response rates (reminders required).
- No control over who answers (spouse, administrative assistant).
- No possibility to aid the respondent, leading to unanswered questions).
- Priming through question order may not work since the respondent can view the entire questionnaire.

Traditional postal surveys are less common as digital solutions take over. Nevertheless, they remain as a relevant way to collect data. Postal surveys require a list of respondents (the sampling frame) and their contact information. Either send questionnaires to everyone on the list, or find a way to take a random sample from the list. Unless it is a topic that the respondents particularly care about, response rates are often very low (around 10%). Response rates above 30% are relatively good. Appeal to them with a good introductory letter, consider offering a reward for answering, and make sure the layout is clear and that it does not take too much time to complete. How much time is too much time is difficult to gauge. The more motivated a respondent is, the more likely they are to invest more time. It is common knowledge that response rates are dropping as people get burned out on answering surveys.

6.10 Summary

In this chapter, we have discussed how to measure constructs, especially in the context of questionnaires. Reliability and validity, and especially construct validity, are essential to assessing the quality of measurements. They evaluate whether the correct construct has been measured, and whether it was measured correctly.

Levels of measurement (nominal, ordinal, interval, and ratio) were discussed in terms of their appropriateness in different circumstances and for how they relate to statistical analysis. We established a fundamental rule of thumb, to always measure at the highest level possible. This is because many of the more advanced parametric statistical methods that we take up later in the book are contingent upon measuring at the interval or ratio levels.

Finally, we spent considerable time on questionnaire design, formulating questions, measurement scale options, and collecting data.

References

Bagozzi, R. P. (1994). In R. P. Bagozzi (Ed.), *Principles of marketing research* (pp. 1–49). Blackwell.

Carmines, E. G., & Zeller, R. A. (1979). *Reliability and validity assessment.* Sage.

Likert, R. (1932). *A technique for the measurement of attitudes*. NY.

Nunnally, J. C. (1978). *Psychometric theory*. McGraw-Hill.

Osgood, C. E., Suci, G. J., & Tannenbaum, P. H. (1957). *The measurement of meaning*. University of Illinois Press.

Sampling

7

Contents

7.1 Introduction

When was the last time you thoroughly and conscientiously answered a survey? Now you find yourself sitting on the other side of the table wondering how to get enough of the right respondents to thoroughly and conscientiously answer your survey. Whether doing qualitative or quantitative research, you need to decide where to get data. The procedure is important for quantitative research because you will run statistical analyses and may want to generalize from the results. It is equally important for qualitative research because you need to identify research objects that can provide rich information on the phenomenon being studied. In this chapter, we will go through different types of samples and how to determine sample size. This includes identifying respondents for qualitative methods like depth interviews and focus groups. We discuss probability and non-probability sampling, and sources of sampling error.

J. E. Sallis et al., *Research Methods and Data Analysis for Business Decisions*, Classroom Companion: Business, https://doi.org/10.1007/978-3-030-84421-9_7

7.2 Define the Population and Sampling Frame

A *unit of analysis* is one member of the group being studied, and it is closely related to the *level of analysis*. When studying biotech companies, each individual biotech company is a unit of analysis, and the level of analysis is at the company level. When studying individual biotech entrepreneurs, then each biotech entrepreneur is a unit of analysis and the level of analysis is at the individual level. A *population* is the sum of all the units being studied. You could, for example, define the population as, "All biotech companies in Sweden." Then, compile a list of all Swedish biotech companies from branch registries or other accessible sources. The list of all the units in a population is called the *sampling frame*.

Imagine doing research on consumer perceptions of the Norwegian Helly Hansen brand in Denmark. To collect data from every person in Denmark is called a *census*. It would be virtually impossible to conduct personal interviews with every person in Denmark. Instead, you might consider doing a questionnaire survey, although it would still be nearly impossible to reach everyone and extremely unlikely they would all answer. An alternative is to select a *sample*, which means to collect data from a relatively small part of the population. Usually, a sample is meant to be *representative* of the population so that, to some degree, results can be *generalized* to the population.

If the research purpose is to find out something about a large group, you are more likely to get better and more accurate information by taking a representative sample than trying to conduct a census. Not taking into account the resources involved in doing a census, if there is a substantial risk that there will be a systematic error in who is contacted and who will answer, the results will be biased. Using the Helly Hansen example, by defining the population as all people over 15 years old registered as living in Denmark, it is practically impossible to contact and collect data from everyone. People without telephones, Internet, email, and permanent addresses are likely to be missed. People who are anti-globalization, anti-free markets, and anti-big business are unlikely to respond. With such a huge amount of data, coding and input errors are likely.

While sampling methods and sample size are more commonly associated with quantitative methods, qualitative research also requires systematic planning when selecting the units of analysis. Deciding whom to recruit is based on logic, not on statistical probability. This could be a certain type of consumer or a particular category of business. Results cannot be generalized in statistical terms. However, if the logic and argument is strong, then the results may be considered representative for the phenomenon being studied. For example, we had students who were doing research on the perceptions of allergy self-test kits for mothers of infants. By specifying demographic parameters like age, income, and education, the students were able to create four different focus group profiles. After the focus groups, the students provided insightful information to the self-test allergy kit producer on key concerns for young mothers. This guided how the company could market their self-test kits to young families.

7.2.1 Sampling Frame

The *sampling frame* is a list of all the units of analysis included in the population, or a description of how the units are identified. Telephone directories, membership lists, and city maps are all examples of possible sampling frames. Telephone numbers associated with a specific geographic area could be used for compiling a sampling frame for doing telephone interviews in that area. A membership list for a loyalty program could provide contact information for doing a questionnaire survey. Choosing sections on a city map could be used for doing door-to-door canvassing. Companies like Facebook have run into legal challenges for selling user information, including contact information, to third parties. People are becoming more sensitive to how their personal information is being used, and many do not appreciate landing on a sampling frame list simply because they have installed an app that requires their contact information.

In practice, the sampling frame may not accurately represent the population (see Fig. 7.1). *Over-coverage* occurs when units of analysis from outside the population are included in the sample. *Under-coverage* occurs when the sampling frame does not accurately list all units of analysis in the population. Lists may not be up to date because people have moved into or out of an area, new companies have started, or old ones have closed, gone bankrupt, or merged. When identifying businesses in industry databases, category definitions may vary between databases, which affects which companies are included in certain categories.

A sample is a subset of the population. In quantitative research, it is common to distinguish between probability and non-probability sampling. A probability sample is characterized by:

(a) That it is possible to determine in advance the probability for each unit to be selected.
(b) That the probability is greater than zero.

All other ways of sampling are considered non-probability samples. There are several subcategories of sampling, of which we explain some of the most common (see Fig. 7.2).

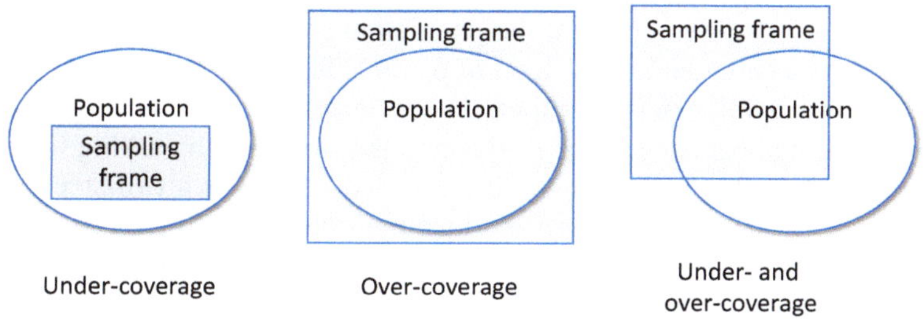

Fig. 7.1 Sampling frame coverage accuracy

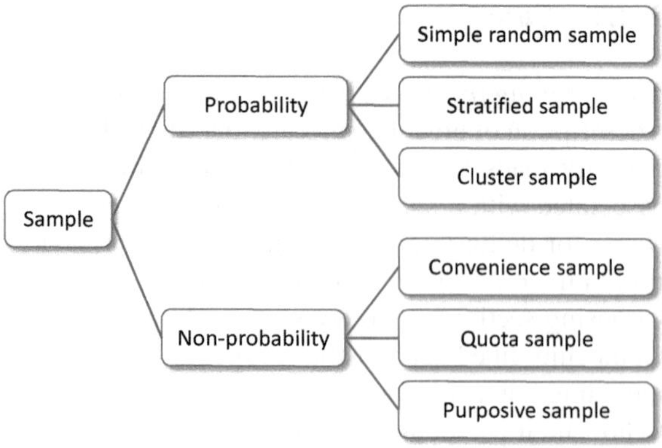

Fig. 7.2 Sample types

7.3 Sampling Method

7.3.1 Probability Samples

Simple Random Sampling
This is the most basic form of probability sampling. To be a simple random sample, it is necessary that:

- Each population unit has the same probability of being selected.
- Each combination of n elements, where n is the sample size, has the same probability of being selected.

The Nordic Viking Lottery provides a classic example of simple random sampling. From a population of 48 balls, six balls are randomly selected from the lottery machine. All the balls, and all the combinations of balls, have the same a priori probability of being selected. The lottery machine containing the balls functions as the sampling frame.

In practice, finding a perfect sampling frame that lists or covers all the units in a population is very difficult. This means that the classic way of drawing a simple random sample, where the population units are numbered from 1 to N, and a random selection of units is sampled from the population, is rare. Instead, try to select units that, as much as possible, represent the population. A typical example is telephone interviewing. In a geographic area, a set of possible household telephone numbers is identified. Then, using CATI (computer aided telephone interviewing), a random selection of the numbers is generated. Often, this is followed by asking for a person over a certain age who had the most recent birthday. This is a two-step procedure where first the household is selected and then the respondent. If the procedure leads to a biased sample, like a predominance of men or women, the selection procedure can be corrected to increase responses from the underrepresented category.

Stratified Sampling

In a stratified sample:

- The population is divided into mutually exclusive and collectively exhaustive subgroups (strata).
- A simple random sample is selected from each subgroup (stratum).
- The sample size in each subgroup may vary.

The main reason for using stratified sampling is to make certain that specific subgroups within the population are represented and that certain variables are sufficiently measured. It is particularly useful when there are heterogeneous subgroups to be represented, and when each subgroup is homogenous for the variables to be measured. With homogeneity within the subgroup, fewer units need to be selected, and weighting can be applied to each of the subgroups to be certain that specific variables will be properly represented at the population level.

Imagine wanting to find out how many cafe lattes the business program students drink in an average week. Also, assume that 60% of the students are male. If you have information suggesting that men drink fewer lattes than women do, you can divide the population into two strata: "female" and "male" students. A simple random sample within each of the strata provides an estimate of average latte consumption for women and men, respectively. Then, calculate average consumption for the population by *weighting* the results of the two strata. If the estimate for female students is 4 lattes per week and the estimate for male students is 2 lattes per week, the population average is 2.8.

$$4 \times 0.4 + 2 \times 0.6 = 2.8$$

The higher the degree of homogeneity in a population, the smaller the required sample size. In the extreme, if all units are identical on a specific variable, then one unit can represent the entire population for that variable. It is this connection between the degree of homogeneity and the sample size that is used when stratifying samples. This also means that the sample size within the individual stratum can vary a great deal, depending on the homogeneity of each stratum. Another reason to stratify may be to guarantee that specific subgroups are represented. With a random sample from the population, there is a risk that important subgroups are not represented. For example, in the Nielsen Norway store index, all large retail chains should be represented. To be certain of this, each retail chain is defined as separate strata, and then the chains are further stratified by store type.

Stratification can take place in several stages and can be very complex. Individual market analysis agencies will have a "master plan" that shows how their stratification and sampling is structured in the standardized surveys they offer.

Cluster Sampling

In cluster sampling:

- The entire population is divided into mutually exclusive and collectively exhaustive strata (clusters).
- A number of strata are randomly selected from the population.
- Within the selected strata, all units can be included in the analysis, or a simple random sample within each strata can be taken.

In a stratified sample, elements from all strata are selected, whereas in cluster sampling, only some clusters (strata) are selected for analysis. All the units of the population must have a known, but not necessarily equal, probability of being included in the sample.

Cluster sampling is best suited to when there is a high degree of heterogeneity within each cluster, and each cluster (stratum) constitutes a miniature of the population. For example, assume that the population is defined as all travelers departing from an international airport during a specified week. Instead of contacting every traveler during the entire week, opening hours for the week could be divided into hourly intervals. Assume the airport is open from 06:00 to 24:00 every day. The interval population is 18 hours × 7 days = 126 hourly intervals, which are called clusters or strata. To be a probability sample, the number of units in each cluster must be known, which in this example would be the number of departing travelers in each hour. The probability of selecting an hourly interval can be set proportionally to the number of travelers in each interval, or let the probability be disproportionate.

For example, randomly draw 10 of the 126 time intervals, and interview all travelers in these intervals or randomly choose every 20th traveler. The point is: all travelers will have a known probability of being selected, making it unnecessary to contact travelers at all times during the week. The problem with clustering is that there can be a large bias if the individual clusters are quite different. With bad luck, all the selected 10-hour intervals may be on Saturday and Sunday, in which case, it is unlikely to get any business travelers in the sample, even though business travelers represent the majority of travelers during the week.

Personal omnibus interviews are often based on a cluster samples. A number of geographical areas are first selected from the population. The probability of selecting an area is often proportional to how many people live in the area. Then, within each area, a random sample is selected. This can be done by defining rules for how the selection will proceed from a given starting point. The advantage is that interviewers do not need to cover the entire geographical area.

Another example of cluster sampling, often called *systematic sampling*, is to start by selecting a random starting point in a sampling frame. For example, by randomly choosing a number between 1 and 100, then from that random starting point, select each 100th unit. The distance (e.g., 1 in 100 or 1 in 1000) is determined by the scope of the list and the desired sample size. For example, if you have a company employee register with 500 names, and you want a sample size of 50, you would randomly choose a number between 1 and 10, and from that starting point select every tenth person. Each group of ten employees is a cluster.

7.3.2 Non-probability Samples

Convenience Sampling

The simplicity of collecting data through social media is very attractive. However, simplicity is not always the best choice. As the label suggests, *convenience samples* are usually achieved by doing what is easiest to accomplish. For example, if the research population is over 15 years old in Oslo, find a crowded place in Oslo and ask anyone over 15 who agrees to answer your questions. The online alternative is to post a questionnaire on platforms like Facebook, Instagram, and WeChat, and collect the responses from anyone who takes the time to answer.

The number of respondents may vary widely, which has nothing to do with being a convenience sample. Even when many people answer the survey, from a statistical perspective, it is not a representative sample from the population. A typical example is when television-viewing audiences are asked to call in and vote for different alternatives being presented in a program. Even though thousands of people may ring in, the results are not statistically representative for the population. Other examples include when questionnaires pop up in web browsers or when you are asked to stay on the line to answer questions after ringing some sort of service like your bank. Only people who are interested participate in the survey, and there is no way to know whether they are representative of the population.

The premise of probability sampling is to be able to calculate how likely (probable) it is that results from the sample deviate from results if the entire population had been measured. With convenience sampling, there are many units (people, companies, etc.) that have no opportunity or intention to be sampled. Consider, for example, television-viewing audiences being asked to call in and vote. Not everyone watches television, not everyone has a telephone, and if it costs to ring in and register a vote, not everyone will be willing to pay. The same thing occurs when intercepting people in public. Busy people, people with small kids, and shy people may decline to participate. People have different habits and preferences, which means they are not all equally likely to be intercepted. If non-participants are systematically similar to each other and they systematically differ from participants, *non-participation bias* occurs.

Quota Sampling

Quota sampling is a slightly more advanced form of convenience sampling. A *quota sample* means that based on information about the distribution of one or more variables in the population, the convenience sample is collected in a way that the distribution matches the population. For example, the population distribution for age and gender may be known. Take Norway as an example. The population distribution for age and gender is accurately known at the country level, and even at the municipality level. If the distribution in the population is 50% men and 50% women, we can sample to have 50% men and 50% women. If we know that in the population, 30% are under 30 years old, 35% between 31 and 50, and 35% 51 and over, we can recreate this distribution in the sample.

The purpose of using quota sampling is that the sample should be more like the population than would be the case if we just took all respondents. An important prerequisite is that the variables on which we base the quotas are important for the phenomena (e.g., opinions, attitudes, perceptions) we are measuring. When it comes to individuals, there is often a high correlation between opinions and age. The same can be said for some topics within gender groups. For businesses, things like firm size and industry affiliation are important. Ultimately, the choice of quota variables is based on the analyst's perception of which variables matter for the distribution in the population for the phenomena being studied.

A quota sample entails that within the limits set by the quota variables, convenience sampling is carried out. For example, if you want a total sample of 200 people, of which 50% should be women, and 30% of both women and men should be under 30 years old, this means recruiting 30 women and 30 men under the age of 30, and 60 women and 60 men over the age of 30. This is done in a convenient way, like through social media or intercept. Nevertheless, even though the distribution of the sample mirrors the population, and thereby is arguably more representative, it is still not a probability sample.

In practice, quota sampling is often used in both qualitative and quantitative studies. In quantitative studies, it is prevalent because of the difficulty, or impossibility, of getting lists of all the elements in a population.

Purposive Sampling

Purposive sampling is the selection of units of analysis based on characteristics of the population that the researcher wants represented in the sample. An example is when "test stores" are chosen for being representative of a retail chain. They are used to test new brands, brand placement, or other marketing initiatives. They have typical properties (localization, sales, customer base, etc.) that make the data representative for particular store types within the retail chain.

In another example, subjects for in-depth interviews are selected to represent different viewpoints in a population. This is done to capture contrasting viewpoints on a phenomenon. The criteria for selection need not be explicit. In Norway, regional managers of a local bank were interviewed, and then asked to provide introductions and contact information for corporate customers who had different types of relationships with the bank. This was done to capture the views of a broad and meaningful cross section of corporate customers. In this case, the composition of the sample was delegated to the regional managers who were the experts in determining the most representative sample. Of course, this assumes that the regional managers made their choices based on attaining the best research outcome, as opposed to political motivations like recruiting customers who they knew would cast them in a positive light.

A hybrid form of purposive sampling that deserves specific attention is called *snowball sampling*. This is often used to recruit respondents who belong to populations that may be otherwise difficult to reach. No sampling frame exists and they are associated through a loosely defined network. You start by identifying a few individuals who belong to the population, and then ask them to identify respondents through their network. The new respondents then identify more respondents. A fitting example comes from a colleague who studied gay consumption habits during the late 1990s. While the gay community was fairly visible, it was, and still is, a challenging population to define and identify. Sexuality and sexual orientation are vehemently debated in the public domain. Definitions are controversial, and there is no sampling frame, no list, of who belongs to the population. Undaunted, he produced superb research on a meaningful group by starting with a few friends, and then through their friends *rolling a snowball* of respondents.

7.4 Sample Size

Estimating sample size is one of the more difficult challenges in business research. It can be broken down into two approaches: (1) non-probability samples determined through pragmatic reasoning and rules of thumb or (2) probability samples determined through stringent statistical reasoning and equations. In either approach, the purpose is to say something about the population from which the sample is taken.

For *non-probability samples*, strictly speaking, there is no statistical basis from which to generalize to the population. Having said this, many statistical techniques require some sort of minimum sample size in order to have the power to statistically detect an effect. *Power* relates to the likelihood of detecting an effect when the effect is present, and can be expressed by the function:

$$Power = f(effect\ size, sample\ size, alpha)$$

By effect, we mean some sort of phenomenon like trust in a relationship. It could be very strong, it could be subtle, or it might not exist. As researchers, we cannot influence the effect size. It is simply the magnitude of the phenomenon we are attempting to measure. Sample size is our greatest tool for influencing power. If the sample size is too small, most likely none of the null hypotheses will be rejected. We explain this in Chap. 9. For now, let it suffice to say that with very small samples we are unlikely to find any interesting relationships in the data. When the sample is too large, then everything appears to be interesting. Alpha is the significance level we set for the hypothesis test. The most usual significance level is 0.05, though it is not uncommon to use 0.1 or 0.01. We explain the significance level in more detail later in this chapter.

Think of power like a magnifying glass, and sample size as the level of magnification. If you look at a mushroom with a 10× magnifying glass, you clearly see the mushroom. If you look at it with 100× magnification in a microscope, you see fiber structures, perhaps some dirt or some bugs. If you use an electron microscope at 500,000× magnification, you might see pollen grains that have fallen onto the mushroom. Sample size, like magnification, plays a role in what you will see in the data. Our advice is to follow the suggested sample sizes that are often connected to particular analysis techniques.

Fig. 7.3 Sample size considerations

Important considerations when determining sample size:

1. The number of groups to be analyzed.
2. Resources (time and money).
3. Proportion of the population with a characteristic.
4. The variance in the population.
5. The required level of precision and confidence.

If you are doing qualitative research that involves interviews or focus groups, you may choose your participants based on logic and theoretical rational. Whom do you think will best provide the data to address your research question? When you start collecting data, much of the information will be new and surprising. As you continue to collect data, there will be fewer surprises. When there is nothing novel coming from new data, you have *saturated the topic* and there is no point in continuing to do more interviews or focus groups. How quickly this happens depends on the homogeneity in the participants with regard to the phenomena you are investigating. On any given topic, this often means 3–4 focus groups or 10–15 interviews.

For *probability samples*, statistical reasoning determines sample size. The equations for calculating sample size are different for different types of samples. Here, we limit ourselves to describing simple random sampling. In Fig. 7.3, we summarize five things that are important to consider when determining sample size. The first three are applicable to any type of sampling, whereas 4 and 5 are specific to probability sampling.

The Number of Groups

If there are no obvious group divisions, then aim for about 100 observations or for the rule of thumb size suggested for the statistical technique you intend to use. These rules of thumb can be found in various sources that focus on the different techniques (e.g., Hair et al., 2014). If the data will be broken down into subgroups, then try to keep the number in each subgroup as balanced as possible and make sure that there are at least 20–50 observations per subgroup. Using rules of thumb presupposes that you have thought through in advance how the data will be analyzed, which depends on the purpose of the analysis. In any case, there will always be people who choose not to respond, so the samples must be large enough to account for non-responses.

Resources

The most pragmatic approach is to consider the total amount of time and money you have to plan, conduct, and report the research. For example, calculate the cost of one interview, and then multiply it to determine how many interviews are feasible to conduct. If you do not have sufficient resources for enough interviews, then you need to reconsider your research design.

Proportion of the Population with a Characteristic

Based on statistical reasoning, the equations for calculating the sample size differ according to which characteristic of the population is in focus. In particular, it is the arithmetic mean and proportion of the population that has a particular property that it is necessary to derive equations for. In practice, it is most common to focus on the proportion of the population that has a characteristic and to determine the sample size based on this. An example might be that we want to estimate how many people have tried a product in the population, and how many have not tried the product.

Variance in the Population

How much your variables vary in the population is crucial for the sample size. At one extreme there is no variation, everyone consumes the same volume or has the same opinion, so there is no need to sample more than one unit. By asking a single respondent, we know what everyone consumes or thinks. The greater the spread (variance) in the population, the larger the required sample under otherwise equal conditions. Most often, we do not know the variance in the population, making direct statistical calculations for sample size redundant.

Precision and Confidence

The purpose of using a sample is to efficiently draw conclusions that can be generalized to a population. The average consumption of a product in a sample is a *parameter estimate* of the average consumption in the population from which the sample was drawn. If we create an interval around the parameter estimate, the width of the interval reflects the degree of precision, expressed as sampling error. *Sampling error* is the difference between the true value and the sample value. If we want to be very certain that the true population value is within the interval, for example, 95% certainty, then we need to have a wide interval. The *degree of confidence* that the true population value is within the interval is called the *confidence level*. Alpha (α) is the proportion outside of the confidence level and is often called the *significance level*. If the confidence level is 95%, then alpha is 5% (also expressed as 0.05). There is a trade-off between precision and confidence level. If we make the interval wide enough, we can be 100% confident that the population value is within the range. However, it is quite meaningless to know, for example, that the average consumption of mineral water per person per week in Norway is between 0 and 30 L.

The norm in social sciences is to set the confidence level at 95%. Increasing the sample size will increase precision. With these two pieces of information, plus the variance in the population, we can derive an equation showing that sample size is a function of (a) the variance in the population, (b) the level of precision, and (c) the confidence level. Most often, the sample is not a large proportion of the population, so the population size has no meaning for determining the sample size. A typical omnibus survey of adults in any Scandinavian country will have a sample size between 1000 and 1500 people. Larger countries may require larger sample sizes; however, this is due to greater heterogeneity in the population, not the population size.

If we focus on estimating the proportion of a population that has a specific characteristic and assuming a single random sample, it may be shown that the sample size (n) can be expressed by the following equation (1):

$$n = \frac{z^2(1 - A)A}{P^2}$$

z: confidence level expressed in standard deviation
A: proportion of the population having the characteristic
P: precision expressed by sampling error

If we want to be 95% confident that the interval includes the proportion of the population with a characteristic, we set $z = 2$. This is because ± 2 (actually 1.96) standard deviations from the mean (average) in a normal distribution include 95% of all observations. In other words, it is only in 5% of cases that we will be so "unlucky" that we get a result that is outside such an interval.

The proportion of the population that has the characteristic (A) is included in the formula; however, the value is unknown (it is what we are trying to find out). A common way to solve this is to assume equal proportions. The term (1 - A)A then reaches its maximum value, which is (1–0.5) 0.5 = 0.25. If we insert these values into equation (1), we get the following equation (2):

$$n = \frac{1}{P^2}$$

If we want to get an estimate of the proportion of the population that has a specific characteristic and set the level of precision (sampling error) at $\pm 5\%$, we put $p = 0.05$ into Eq. (2). In this example, $n = 400$. We need a sample of 400 individuals to be sure that the level of precision is $\pm 5\%$. If the proportion having the property is higher than 50%, the level of precision will increase.

For any given sample size, sampling error increases as the proportions become more even. It is also worth noting that the sample size must increase quite a bit in order to achieve a meaningful reduction in sampling error (or increase in precision) when the sample size exceeds a certain level. For example, with a 50% proportion, the sample size must be quadrupled, from 400 to 1600, in order to halve the sampling error from 5% to 2.5%. Since data collection costs increase more or less proportionally, such large samples are rare.

It is important to remember that the sample sizes we are discussing refer to the final sample size after taking into account non-responses. In other words, if we expect to achieve a response rate of 50%, then the gross sample must be twice as large as what we calculate using the equations. A response rate of 30–35% is considered quite good for questionnaire surveys, which means that the gross sample should be tripled in relation to the number indicated by purely statistical reasoning. It is convention to report the non-response rate, which is the non-responses divided by the sample size.

7.5 Error Sources

Though it has broader relevance, our discussion in this section pertains primarily to errors when collecting data through questionnaires and interviews. Error sources can be divided into two main categories: missing observations and measurement error (see Fig. 7.4).

7.5.1 Missing Observations

We will discuss three types of missing observations: coverage errors, non-response errors, and sampling error.

Coverage Error
Coverage error occurs when the sampling frame either includes units that are not in the population (over-coverage) or does not include all the units in the population (under-coverage). This occurs when things like membership lists are not up to date or category definitions are wrong. We discussed this in more detail earlier in the chapter around Fig. 7.1, sampling frame coverage accuracy.

Non-response Error
Non-response error is a result of sample units not providing data. For questionnaires, this means that some people who were sampled did not complete and return the questionnaire. This is a common type of error, which is why it is important to take a large enough sample to account for non-responses. It also occurs when taking a census, meaning we attempt to collect data from the entire population. So long as the error is *random*, the consequence for the research is that the parameter estimates will be less precise. In other words, the confidence intervals will need to be wider, although the estimates themselves are considered correct. When the error is

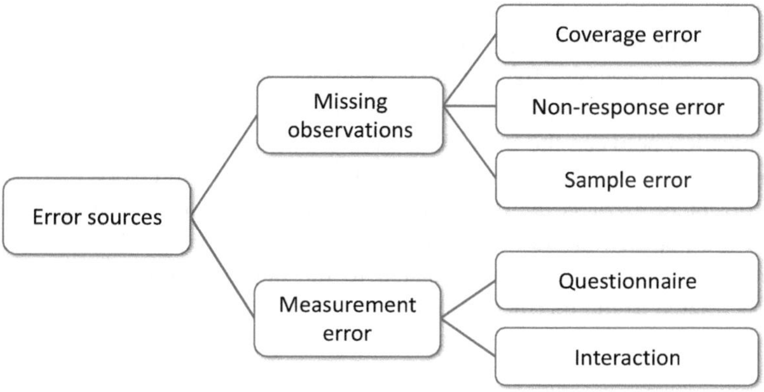

Fig. 7.4 Error sources

systematic, meaning it is not random, the consequences can be quite serious because the parameter estimates are biased in one direction or the other, which in practical terms means they are systematically incorrect. The response rate provides a measure of the severity of non-response error. However, it is not the response rate, per se, that indicates a problem. The problem is whether, regardless of the response rate, an important group in the population is not represented in the sample. Response rates are discussed in more detail in the section on modes of communication (mail, telephone, personal interview).

In addition to the response rate, to evaluate the quality of the sample, it is important to know some background information about the population: things like the distribution with respect to sociodemographic variables such as gender and age. You test the sample to evaluate whether the distribution of respondents on background variables differs significantly from the distribution in the population. Similar distribution indicates that non-response error is limited. However, when the research variables (e.g., behavior and attitudes) are not clearly linked to the background variables, such as age and gender, such an analysis is not definitive evidence of low non-response error.

Sampling Error

We had a comprehensive discussion of sampling error earlier in this chapter under the section on precision and confidence. *Sampling error* is the difference between the true value in the population and the sample value. It is important to consider when we have taken a probability sample and then generalize findings to the population. The calculation of sampling error is based on statistical theory and does not account for coverage error and non-response error. All of these errors can result in incorrect and potentially biased parameter estimates. Again, this is a matter of whether the errors are random or systematic.

Measurement Error

The other main category is measurement error. It occurs when respondents answer the questionnaire, but for some reason the answers do not correctly measure the variable. The purpose of the questionnaire is to measure "true" characteristics of the respondent, whether these are background variables such as age, gender, and income, or other variables like attitudes, opinions, behaviors, and perceptions. The starting point is that questions and answers in a questionnaire are a form of communication between the interviewer and the interviewee. As in all communication, there is room for many kinds of misunderstandings and errors.

Measurement error is the difference between the measured values and the true value in the population, and it can be random or systematic. We can distinguish between errors that are primarily related to the questionnaire and errors that are primarily related to the interaction between the interviewer and the respondent.

Questionnaire

Errors that are primarily related to the questionnaire are possible to limit by carefully designing and pre-testing the questionnaire. We discussed this in detail in Chap. 3, in

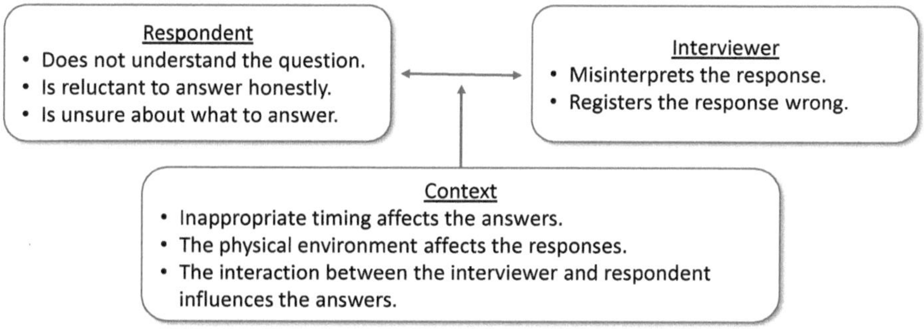

Fig. 7.5 Sources of error from interaction

the section on survey research with questionnaires. Besides pre-testing, the other fundamental piece of advice is to always begin by looking at existing measures. In the natural sciences, researchers do not invent a new temperature scale whenever they want to measure temperature. In the social sciences, we should refine existing measures, not continually reinvent measures.

Interaction

Errors related to the interaction between the interviewer and the respondent can be mitigated by selecting suitable interviewers and taking the time to prepare and train them. In Fig. 7.5, we have listed some typical errors related to the interaction between the interviewer and the respondent.

The causes of measurement errors shown in Fig. 7.5 are quite intuitive. For example, most people who have received a telephone call from a marketing research company know that such calls are often inconvenient. Nevertheless, if you choose to participate, the stress you feel for the time it takes can affect what you answer. Without the stress and with time to reflect, you may answer quite differently. Research shows that the physical surroundings affect interview and questionnaire responses. Noisy contexts can be distracting or perhaps something is drawing your attention away from the questions. The importance of the interaction between interviewer and respondent lies in the fact that people behave differently in different social situations. Some will give answers that they think will impress or please the interviewer, if it is a person the respondent wants to impress. The interview situation can be a "game," where the respondent may be more concerned with surprising the interviewer than with giving truthful answers.

Different respondents may have different response styles. By style, we mean that respondents have different approaches to answering questionnaire surveys. For example, Harzing et al. (2012) investigated questionnaire response style differences between national cultures. They compared an *extreme response style*, where the respondent is inclined to use the extreme poles on the response scales, versus a *middle response style*, where the respondents tend to use the centers of the scales and avoid extreme answers.

They found that Eastern cultures tend toward a middle response style, whereas Western cultures tend toward an extreme response style. The response styles may be a more or less permanent characteristic of a respondent; however, they may also arise as a result of the interview situation and the topic. They found that 7-point scales (like we recommend) reduce the effect of response style bias when compared with 5-point scales. They also found that differences were less when experts answered the questionnaire. We recommend to always recruit knowledgeable respondents. When respondents are not familiar with a topic they tend to guess, which adds error to the responses. In large studies involving different types of people, the impact of this type of measurement error is less severe. The various errors balance each other out. If the number of respondents is limited and consists of homogeneous individuals, a given response style may be dominant, thus introducing a systematic error.

7.6 Summary

In this chapter, we have gone through different types of sampling methods and discussed procedures for determining sample size. We discussed populations and sampling frames. Probability sampling is necessary for generalizing based on statistical theory. We discussed simple random sampling, stratified sampling, and cluster sampling. Non-probability sampling is common in both qualitative and quantitative research. Generalization of results is based on logic and argumentation, and has no direct basis in statistical theory. We also went through ways to determine sample size.

References

Hair, J. F., Black, W. C., Babin, B. J. & Anderson, R. E. (2014) *Multivariate Data Analysis, 7th edition,* International edition, Pearson Education Limited.

Harzing, A.-W., Brown, M., Köster, K., & Zhao, S. (2012). Response style differences in cross-national research: Dispositional and situational determinants. *Management International Review, 52*(3), 341–363.

Part III

Quantitative Data Analysis

Simple Analysis Techniques

8

Contents

8.1 Introduction

A relatively recent data analysis trend is to examine clickstreams on the Internet: which pages you visit, how long, what you click on, the frequency of clicks, which pages you came from, and which pages you go to. Clickstream analyses look at correlations and patterns in online behavior. Take note, clickstreams do not say anything about the motives behind the behavior or who the person is. Nevertheless, the data is so extensive that the patterns still provide valuable information. In this section of the book, we will learn about the most common quantitative analysis methods. Our examples are done in IBM SPSS Statistics software. Basic statistics can be done in Excel, although it has the disadvantage of not being a dedicated statistics software so it is, relatively speaking, quite limited. Other popular software are JMP, SAS, and Stata. Each has its own advantages and disadvantages. Often, different scientific domains or even geographic areas favor different software. It is wise to consider where and with whom you will be working before deciding on which software to use. One important distinguishing characteristic is whether a software has a windows-based point and click interface or whether it requires syntax-based programming. While point and click may be intuitively easier, syntax-based programming has the advantage of leaving a clear trail of how the

J. E. Sallis et al., *Research Methods and Data Analysis for Business Decisions*,
Classroom Companion: Business, https://doi.org/10.1007/978-3-030-84421-9_8

analysis was conducted. One does not always remember what choices were made in the point and click environment. SPSS is a windows-based point and click software, though it can be programmed with syntax. To leave a trail of what choices were made, when in an SPSS dialogue box, clicking on "paste" will generate a syntax file of the requested analysis. On the book's website, you can get the data for the examples in each software format.

8.2 Using Software

When you open SPSS, you see a dialogue box over top of a spreadsheet. You can choose to create a new dataset or open a recently used file. By simply closing the box, you have a blank spreadsheet where you can create a new dataset. The variables and data can come from many sources. We show an example of inputting data from a questionnaire survey. You choose *Variable View* at the bottom left (see Fig. 8.1).

The next step is to input the variables (see Fig. 8.2):

1. First, give each variable a *Name*, which in our example is gender.
2. Leave *Type* as numeric and set *Decimals* to 0.
3. Click on *Values* to open a dialogue box. Input the appropriate values.
4. Set the appropriate *Measure level*, which for gender is *Nominal.*

You can add a label, which is often a more detailed name.

In Fig. 8.3, we have input data for seven respondents for three variables (Gender, Age, and Attitude). We show the *Data View.*

Notice that Attitude for respondent 4 is empty. We call this *missing data.* When left blank, the program will treat it as missing. We can also define missing values as a number we choose, and then input that number. When choosing a value for missing

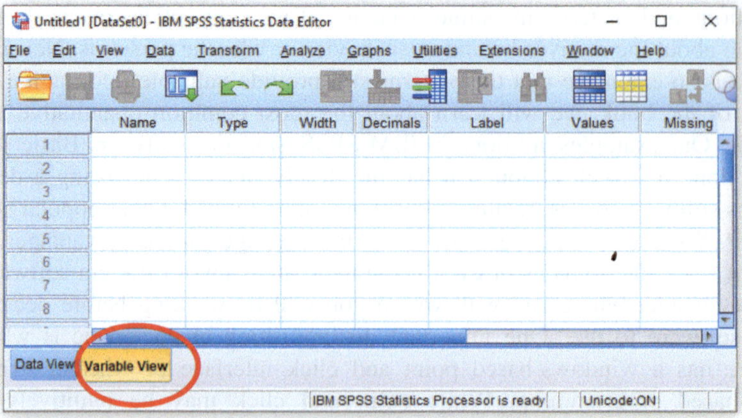

Fig. 8.1 Variable view in SPSS

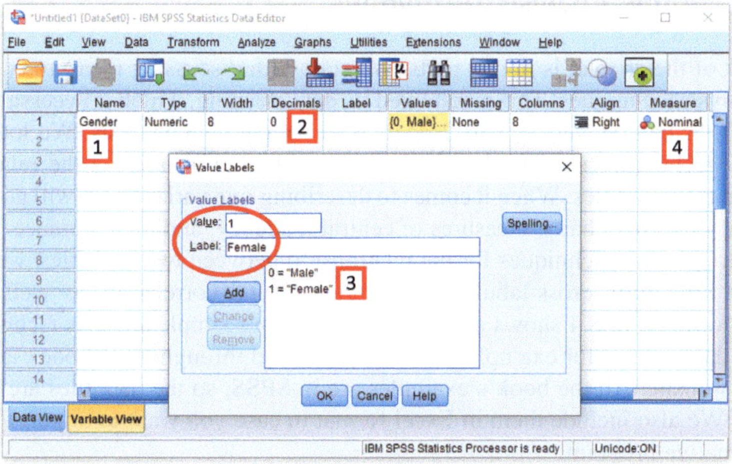

Fig. 8.2 Inputting variables

Fig. 8.3 Inputting data

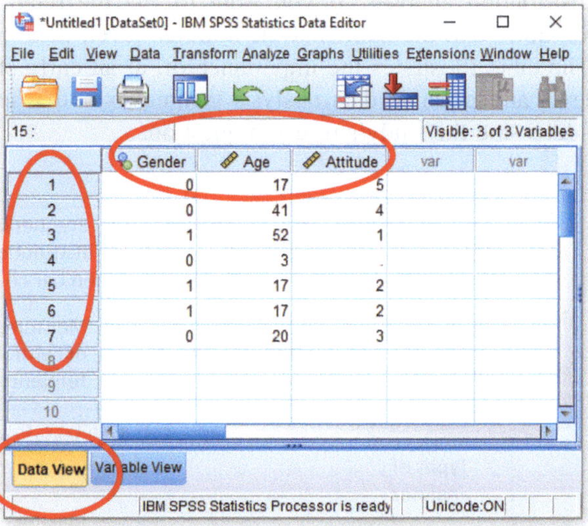

data, as a rule of thumb, you should pick an absurd value relative to your data. Typically, something like −99 for questionnaire data. This is so that missing values are easily identified, and if you forget to define −99 as missing, your results will be so strange that, hopefully, you will quickly identify your mistake.

8.3 Simple Analysis Techniques

The rest of the chapter is in two parts. We start with discussing measurement levels and analysis techniques. This is followed by data cleaning, which is necessary before the analysis can be started. Part 2 deals with simple data analysis. We look at the analysis techniques used to describe one variable and to analyze the relationship between two variables. When it comes to describing one variable, we will go through frequency and bar charts, measures of central tendency, and then variance. When it comes to analysis techniques for the relationship between two variables, we will go through reliability, cross-tabulation, covariance, and correlation, as well as rank correlation. Figure 8.4 shows an overview of various simple analysis techniques.

The data we use for examples in this chapter and throughout the book are on the book's website. All the book's examples are in SPSS, so the data files are in SPSS format. We also include them in Excel format in case you want to import them into an alternative program.

Table 8.1 is a guide for which analytical techniques can be used depending on the scale level and the number of variables to be analyzed. For example, the table shows that when you have a nominal scale and one variable, you can use frequency distribution analysis. If you want to test the relationship between two or more variables and the data is at the nominal level, then cross-tabulation is appropriate. The table illustrates the importance of, at an early stage, thinking about the scale level and possible data analysis techniques and whether they are consistent with addressing the study's research question.

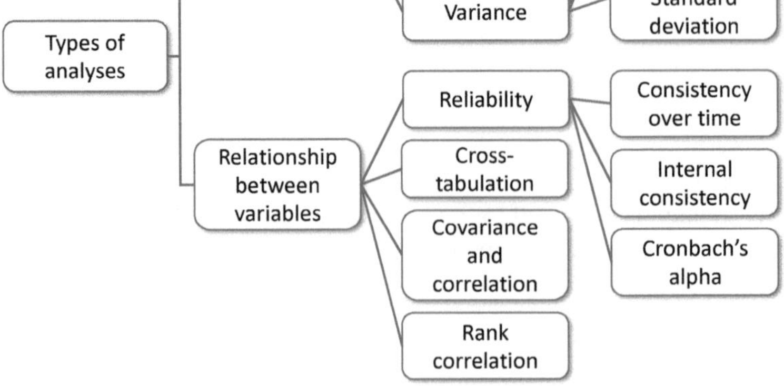

Fig. 8.4 Simple analysis techniques

Table 8.1 Scale level and analysis technique

Scale level	One variable	Relationship between variables
Nominal	Frequency	Cross-tabulation
Ordinal	Median, rank, frequency	Rank correlation
Interval	Mean, variance & standard deviation	Covariance & correlation
Ratio	Mean, variance & standard deviation	Covariance & correlation

In Chap. 6, we discussed how statistical analyses can be divided into nonparametric methods for nominal and ordinal level data and parametric methods for interval and ratio level data. There is a much wider array of parametric methods, and higher-level measures can be transformed to lower levels, though not the other way around. Therefore, our advice is to always measure at the highest level possible. Nevertheless, if the measures are at the nominal or ordinal level, keep in mind that there are many nonparametric methods.

8.4 Cleaning the Data

Before embarking on the actual data analysis, you should inspect the data for coding errors, outliers, and missing values. *Coding errors* are illogical values in the dataset. If a variable can only contain values from 1 to 7, then all values outside this range are coding errors. *Outliers* are *extreme values* that deviate substantially from what is typical for the variable. If a variable can assume values between 1 and 100, and the majority of observations (80–90%) are grouped from 1 to 20, then values like 70, 80, or 90 should be considered as outliers. Outliers have adverse effects on many statistical estimates. For example, the mean of a variable is quite sensitive to outliers. It is the same for regression analysis with ordinary least squares (OLS) estimation, which we cover later in the book.

Coding errors and outliers are detected by thoroughly going through the data. Graphical representations (e.g., frequency distributions) and various descriptive statistics are very helpful in finding them. When possible, coding errors should be replaced with the correct value from the data source. Imagine that you have input data from questionnaires where the values range between 1 and 7. In the dataset, you discover values like 55 and 66. Likely, they were meant to be 5 and 6, respectively. However, you cannot be certain. Go back to the source—the questionnaire for that respondent—and check!

If an outlier is not a coding error, what should you do with it? The answer depends on what kind of analysis you intend to perform. If you are only showing frequency tables and distributions, you do not need to do anything. Outliers do not influence the analysis; they are simply part of the data presentation. On the other hand, if more advanced estimation techniques are used, you must at least be aware of the adverse effects that outliers can have on the results. If you are in the fortunate situation of having a large dataset and the number of outliers is relatively small, you should probably consider removing the cases where the outliers occur. However, it should be noted that there is no formula or rule governing how to address outliers. Common sense and argumentation play an important role. For example, in Scandinavia there are an abundance of SMEs (small to medium-sized enterprises), and some huge multinational enterprises. Imagine that you are doing research on organizational processes (how companies function) in Scandinavia. Most likely, huge multinationals function quite differently than SMEs. If you include all Scandinavian companies in the dataset, the relatively few huge multinationals are likely to have outliers on many variables. By excluding them from the analysis, you will get representative results for Scandinavian SMEs. We refer you to Aguinis, Gottfredson and Joo (2013) for a detailed discussion of outliers.

Missing values occur when there is no observation recorded in one or more cells in a dataset. They leave holes that have not been assigned a number value. When data comes from questionnaires, missing values are simply due to the respondent not answering all the questions or an error in data input. In SPSS (this is common in statistical software), they appear as an empty cell with a dot in it. There can be many reasons for missing data, which we will not go into here. In Chap. 7, we discussed the importance of determining whether data are missing at random or whether there is a systematic pattern to why they are missing. Assuming they are missing at random, it is important to consider how to treat the missing values. This is a complicated topic and too comprehensive for us to go into detail here. However, we will offer some advice. There are three basic possibilities (though not complete solutions):

1. Listwise deletion, which is to omit all cases with missing values. This works well when there is very little missing data relative to the sample size.
2. Pairwise deletion retains more data than listwise by using cases where data is available for each pair of variables in the analysis. Where listwise removes all cases with any missing data, pairwise only removes cases when the data is missing between each pair of variables being analyzed.
3. Impute (replace) missing values with a neutral value. There are several types of imputation. For example, the neutral value may be the average of the non-missing observations in the variable or it may be based on a pattern present in the data.

These options, and others, are automated in the statistical software. The best advice is to, as much as possible, avoid getting missing data. So long as it is minimal (e.g., under 10% for any given variable), then how you treat it is relatively unimportant. The more missing data you have, the more important it is to understand the rammifications of how you treat it.

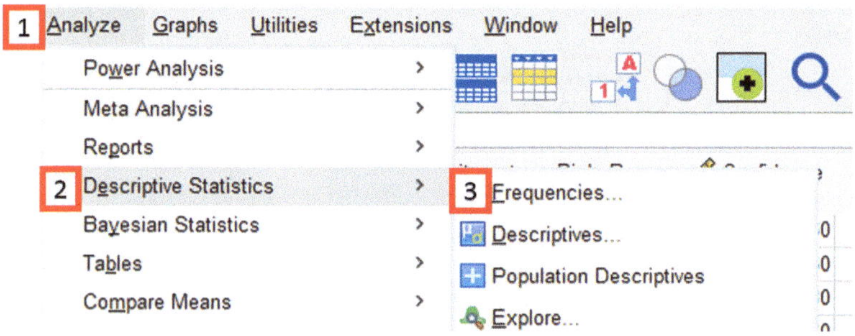

Fig. 8.5 Frequency analysis

8.5 Analytical Techniques for One Variable

In this section, we use the following notation:

f	Frequency
X	Variable X
$\overline{X}$	Generic notation for the sample mean, or sample mean of variable X
Y	Variable Y
$\overline{Y}$	Sample mean of variable Y
$Cov\ (X,Y)$	Covariance between variable X and variable Y
$Corr\ (X,Y)$	Correlation between variable X and variable Y
n	Sample size
σ^2	Population variance
σ	Population standard deviation
s_X^2	Sample variance for variable X
s_x	Sample standard deviation for variable X

Frequency Distribution and Bar Charts

Often, you need an efficient presentation of the data. If the dataset is small and clear, comprehending the information in the data is relatively easy and quick. If, on the other hand, the data material is large and unclear, such as the *Hotel* data on the book's website, frequency distributions and graphical representations (e.g., bar charts or histograms) can be of great help. For a frequency distribution, one simply counts how many times the individual observation values are repeated. To calculate the relative frequency, divide the frequency number (f) by the sample size (n). To follow our example, in SPSS, open the *Hotel* data, and choose: *Analyze > Descriptive Statistics > Frequencies* (see Fig. 8.5).

Then, *choose Month_open, and move it to the Variable(s) box*. The Month_open variable represents how many months per year the participating hotels are open. Choose: *Charts > Bar charts > Continue > OK* (see Fig. 8.6).

The output is in Table 8.2. At the bottom of the table, you can see that there are 100 observations, which include 2 missing values. Missing values are common

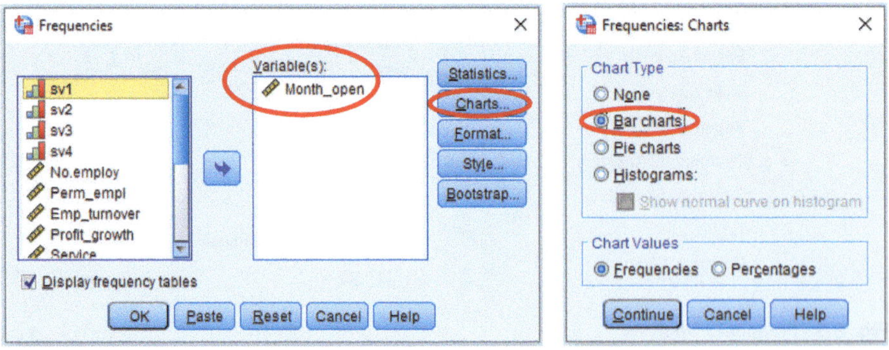

Fig. 8.6 Frequency analysis 2

Table 8.2 Frequency results for months open

Month_open

		Frequency	Percent	Valid Percent	Cumulative Percent
Valid	5	1	1,0	1,0	1,0
	6	1	1,0	1,0	2,0
	9	1	1,0	1,0	3,1
	10	6	6,0	6,1	9,2
	11	7	7,0	7,1	16,3
	12	82	82,0	83,7	100,0
	Total	98	98,0	100,0	
Missing	-999	2	2,0		
Total		100	100,0		

when using questionnaires because not all the respondents answer all the questions. The Frequency column shows the number of observations in each category, which in this example are the 12 months of the year. The Percent column shows the frequency in each category converted to percent. The vast majority, 82 hotels, which represent 82% of the sample, are open year-round for 12 months. One hotel is open for 5 months.

The same data can be presented as a bar chart (see Fig. 8.7). To get the data labels, in SPSS *double click on the chart.* When it opens in another window, *click on Elements, and then, Show data labels. Close the window.* You can quickly see on the bar chart that 82 hotels are open for 12 months, whereas 7 for 11 months, 6 for 10 months, and so on.

When data are at the nominal or ordinal level, reporting frequency tables and bar charts is common. When data is at the interval or ratio levels, there are often too many categories for frequency tables and bar charts to make sense. Do a frequency

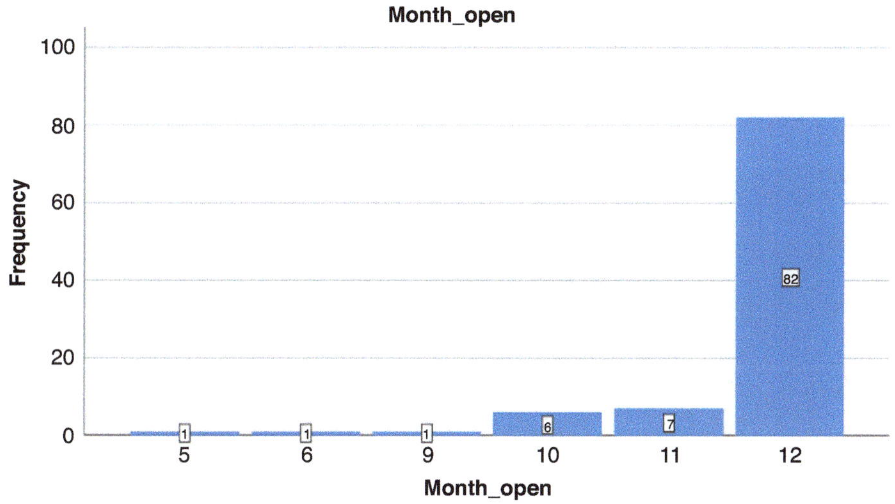

Fig. 8.7 Bar chart for months open

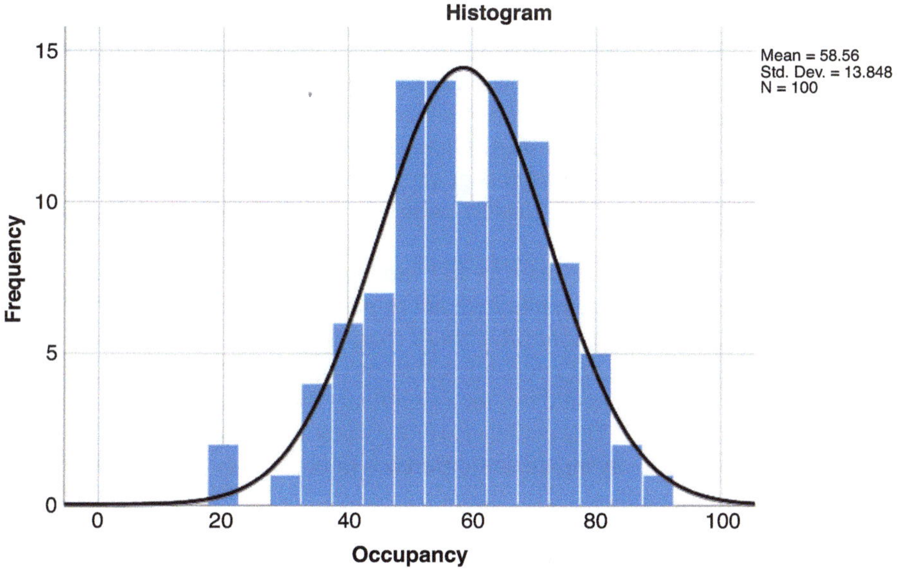

Fig. 8.8 Histogram of occupancy data with normal curve

analysis on the Occupancy variable, but this time in the Frequency dialogue box, choose: *Charts > Histogram > click the box Show normal curve on histogram > Continue > OK.*

Though we do not show it here, if you are following the example in SPSS, you will see that frequency tables can be so large that they are difficult to interpret. In

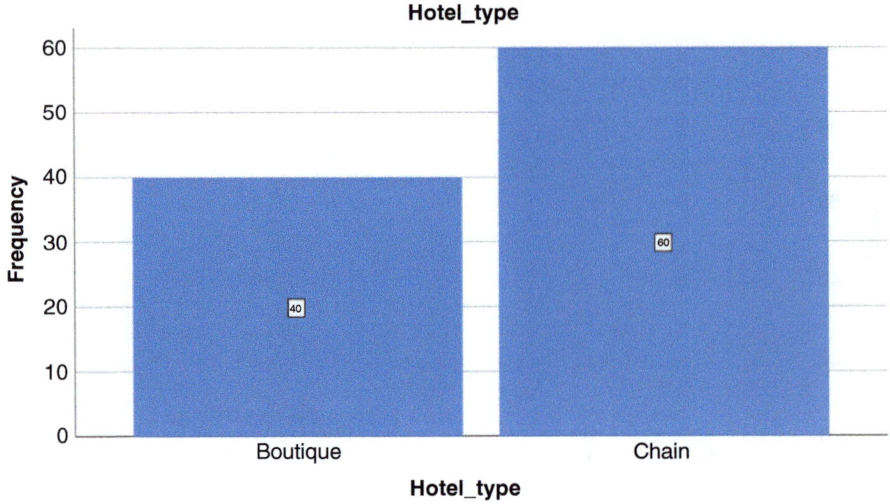

Fig. 8.9 Bar chart of Hotel type

Fig. 8.8, we show the histogram with a normal curve imposed on it. The average occupancy rate is about 60%, which we can see by reading the mean (58.56) and looking at where the distribution peaks. The diagram provides a quick impression of how the data is distributed. There is no universally accepted single measure saying whether data is normally distributed. Often, it is a combination of statistics and visual representation. Histograms change depending on how the axes are defined. Therefore, always check at least a few measures and graphs when assessing the distribution.

In Fig. 8.9, we show a bar chart for the variable Chain, which tells whether the specific hotel is a stand-alone boutique hotel or part of a branded chain of hotels.

We can see that there are somewhat more chain hotels (60) than boutique hotels (40) in the sample. Later in the book, we will show how to test for group differences. In our current example, we could test for a significant difference between hotel types and their occupancy level. This could answer the question, assuming everything else equal, is there a statistically significant difference in occupancy rates for the type of hotel (chain or boutique)?

Measures of Central Tendency

Measures of central tendency are used to show the middle of a distribution of data. The two measures we will go through are the *mean* and the *median*.

(a) Mean

The most popular measure of central tendency is the average, or what statisticians refer to as the *mean*. The notation for the mean is to put the variable identifier with a line on top. For variable *X*, the sample mean is, for variable *Y*, the sample mean is, and so on. If you have a set of numbers (data), you can calculate the mean according to the formula below. The mean indicates the center of gravity in the data. However, the mean has a number of shortcomings. If the data contains outliers, the mean can be greatly influenced in one direction or the other. For example, we saw in the *Hotel* data that the vast majority of hotels (82%) are open year-round for 12 months (see Table 8.2 and Fig. 8.7). Reporting the mean, which is 11.64, is misleading. Only reporting the mean for data with outliers can hide the true nature of the data. The following simple example should illustrate this: The age of everyone present at a birthday party was 6, 7, 4, 6, 3, 4, 6, 5, and 85 years old. There were eight children and one grandparent. The average age of everyone there was:

$$\overline{X} = \frac{3+4+4+5+6+6+6+7+85}{9} = 14$$

We sum the *X* values and divide by the number of n observations to get the mean value of 14. Without knowing what lies behind the data, you would expect to see a group of teenagers. The grandparent outlier pulled the mean up, so much so as to misrepresent the data. The general formula for calculating the average can be displayed as follows:

$$\overline{X} = \frac{X_1 + X_2 + X_3 + \ldots + X_n}{n} = \frac{1}{n}\sum_{i=1}^{n} X_i$$

The mean, strictly speaking, should only be calculated for data at the interval or ratio level. For example, if you have a variable that differentiates between different colors and you choose to code red = 1, yellow = 2, green = 3, and blue = 4, it is pointless to calculate the average value. What color would 3.4 be? It is also theoretically wrong to calculate the mean for ordinal data. Here, however, it is common to compromise by assuming equal distances between each level on the ordinal scale. Typically, this is done for the Likert-type scales that we discussed in Chap. 6. The equal distance assumption is borrowed from interval level scales and allows for the use of many statistical methods developed for data at the interval and relationship level.

We often want data to have a central tendency whereby observations gather symmetrically around the mean. The Occupancy data in Fig. 8.8 follows a bell-shaped curve around the mean. The data for the kid's birthday party does not have a central tendency. It has a clump at one end of the scale with the kids and at the other end with the grandparent. Another example is tea consumption. People polarize around preferring ice tea and hot tea, with very few drinking room temperature tea. Nevertheless, the average temperature for tea consumption is probably close to room temperature.

(b) Median

Another widely used measure of central tendency is the median. The median does not have the same important statistical properties associated with the mean; however, sometimes it is a more appropriate measure. The *median* is defined as the middle

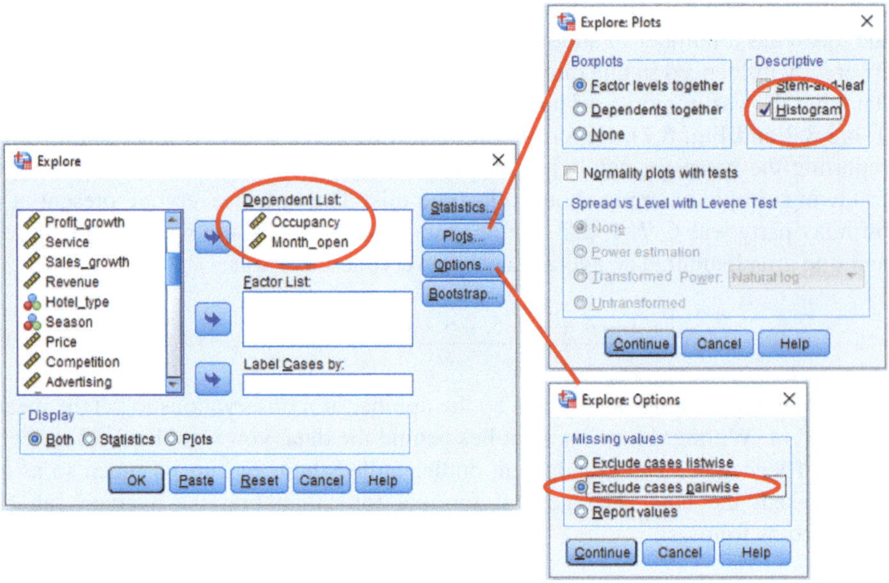

Fig. 8.10 Descriptive statistics: Explore

observation when all observations are arranged in an ascending order. For the children's party, the middle number is 6, which is the median. In this case, the median appears to be the appropriate measure for the age composition of the birthday party. If the variable has an even number of observations (there is no middle observation), then the median is the average of the two middle observations. For example, if we exclude the grandparent from the birthday age data, there are 8 observations. The average of the middle fourth and fifth observations is:

Children's ages = 3, 4, 4, 5, 6, 6, 6, 7

$$\text{Median} = \frac{5+6}{2} = 5.5$$

We will now show how we calculate the mean and median for some of the variables in the *Hotel* data. We use how many months per year the hotel is open (Month_open), and the occupancy rate (Occupancy). In SPSS, choose: *Analyze > Descriptive Statistics > Explore.*

Move *Occupancy* and *Month_open* to the *Dependent List*. Click on: *Plots > Histogram > Continue > Options > choose Exclude cases pairwise > Continue > OK* (see Fig. 8.10).

As we can see in Fig. 8.11, in the *Case Processing Summary* at the top left, all hotels have reported occupancy rates and 98 of 100 hotels have reported their months open. The 2 hotels that did not report are considered missing values. It

Case Processing Summary

	Cases					
	Valid		Missing		Total	
	N	Percent	N	Percent	N	Percent
Occupancy	100	100,0%	0	0,0%	100	100,0%
Month_open	98	98,0%	2	2,0%	100	100,0%

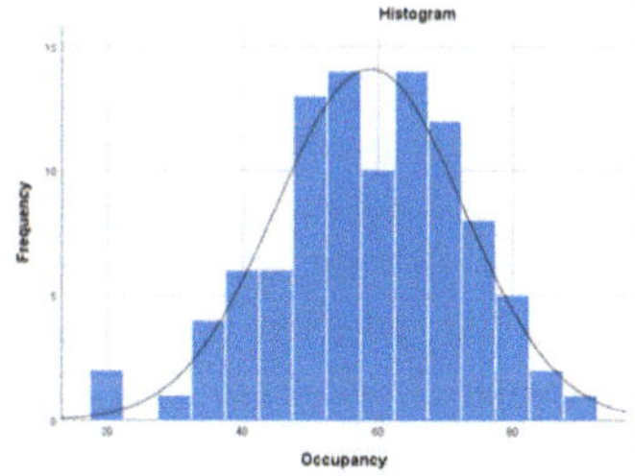

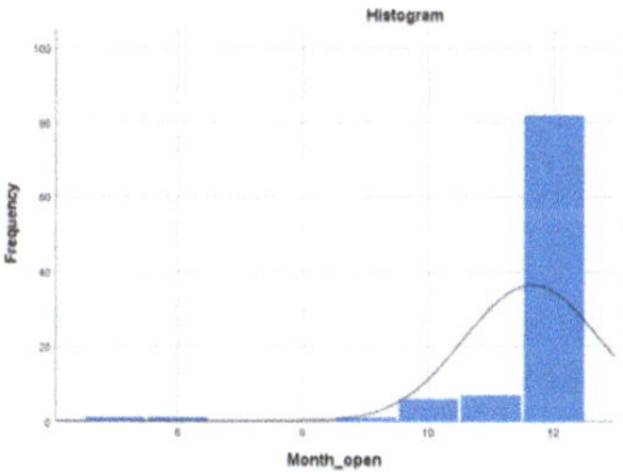

Descriptives

			Statistic	Std. Error
Occupancy	Mean		58,56	1,385
	95% Confidence Interval for Mean	Lower Bound	55,81	
		Upper Bound	61,31	
	5% Trimmed Mean		58,84	
	Median		60,00	
	Variance		191,764	
	Std. Deviation		13,848	
	Minimum		20	
	Maximum		88	
	Range		68	
	Interquartile Range		20	
	Skewness		-,323	,241
	Kurtosis		-,026	,478
Month_open	Mean		11,64	,109
	95% Confidence Interval for Mean	Lower Bound	11,43	
		Upper Bound	11,86	
	5% Trimmed Mean		11,83	
	Median		12,00	
	Variance		1,160	
	Std. Deviation		1,077	
	Minimum		5	
	Maximum		12	
	Range		7	
	Interquartile Range		0	
	Skewness		-4,348	,244
	Kurtosis		21,823	,483

Fig. 8.11 Descriptive statistics for Occupancy and Months Open

may have been data that was difficult for the respondent to provide, or possibly they consider it private and of strategic importance. We see that the *Occupancy* mean (58.56) and median (60.00) are quite similar. This is also reflected in the bell-shaped curve of the frequency distribution that we can see in the histogram. Compared to the very skewed distribution of the *Month_open* variable, the mean (11.64) suggests that the average hotel is open somewhere around 11 months per year, whereas we know from the frequency analysis that 82 of 100 hotels are actually open 12 months per year. The median (12.00) gives a more accurate description of the skewed variable.

Skewness and Kurtosis

In Fig. 8.11, there are two sets of numbers bordered in green: skewness and kurtosis. They express how much the empirical (observed) distribution deviates from the theoretical bell-shaped normal distribution. The *skewness* measures how far away an observed distribution is from a theoretical symmetrical distribution. The excess *kurtosis* is a measure of how peaked or flat the distribution is compared to a theoretical normal distribution. If the distribution is symmetrical, then the skewness is 0, and if the distribution has the same kurtosis as a normal distribution, then it is called mesokurtic and the excess kurtosis is 0. The accepted cutoff indicating a normal distribution in SPSS, and all statistical software that we are aware of, is an absolute value of 1.

Table 8.3 Interquartile range

Q1		Q2	Q3
25%	**25%**	**25%**	**25%**
lowest–highest	lowest–highest	lowest–highest	lowest–highest
Interquartile range = Q3–Q1			
Median = Q2			

A negative skew means the data leans to the right, whereas a positive skew means the data leans to the left. This is easily demonstrated by looking at the histogram for the Month_open variable in Fig. 8.11, and comparing it with the skewness of −4.348. When kurtosis is negative, the distribution is flat; when it is increasingly positive, it becomes peaked. The Month_open variable has a kurtosis of 21.823 and a clearly peaked histogram. The Occupancy variable has a quite normal histogram with a skewness of −0.323 and a kurtosis of 0.026. Our conclusion is that Occupancy is normally distributed, whereas Month_open is very abnormally distributed. Month_open is problematic for parametric statistical methods that require a normal distribution.

Variance

The different measures of central tendency do not always provide sufficient information about a dataset. We also want to say something about the spread of the data. If we return to the *Hotel* data, we see that there is variation in how respondents answer questions. *Variance* is a measure of how the data is spread out around the mean. This is partly because respondents think differently about specific questions, and partly due to respondent error. There are various measurements of spread.

(a) Range and Interquartile Range

The *range* is defined as the difference between the largest and smallest values in a variable. In the children's birthday example, the range is 85–3 = 82. Another measure of the spread, which is not so sensitive to extreme values, is the *interquartile range*. It is defined as the difference between the third quartile and the first quartile. Quartile has to do with 4: You divide the ordered number material into four equal groups (if you have 100 observations, there will be 25 in each group). The smallest numbers are in group 1, the second smallest in group 2, and groups 3 and 4 contain the second largest and largest numbers.

In Table 8.3, the largest number in group 1 is called the *first quartile*, and the largest number in group 3 is called the *third quartile*. The difference between these two is the *interquartile range*. The bottom and top 25% of the data are left out, thus reducing the influence of outliers. In SPSS, boxplots show the median and interquartile range, as well as outliers if there are any (see Fig. 8.12). In SPSS, outliers are cases with values between 1.5 and 3 times the interquartile range. Extreme outliers are above 3 times the interquartile range. When requested, boxplots are generated in Descriptive Statistics, Explore; or in Chart Builder. The symmetry

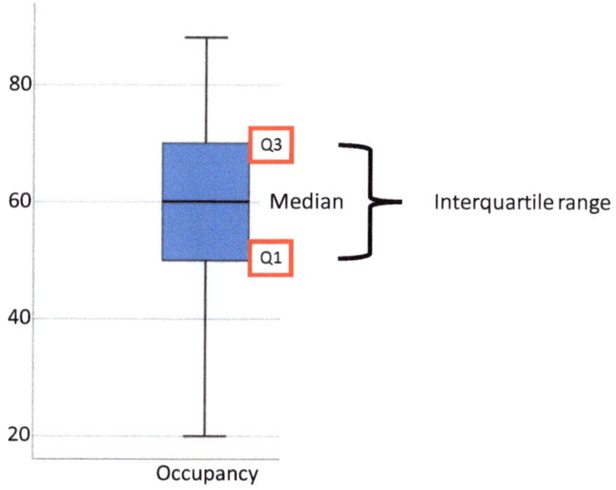

Fig. 8.12 Boxplot

Table 8.4 Identical means of tea

Descriptive statistics					
	N	Minimum	Maximum	Mean	Std. deviation
Tea1	100	1.00	10.00	5.6000	2.18350
Tea2	100	1.00	10.00	5.6000	3.43188
Valid N (list wise)	100				

of the boxplot gives an idea of the data distribution. Perfect symmetry means a normal distribution.

(b) Variance and Standard Deviation

Sample variance s_x^2 and standard deviation s_x are far more important than range and interquartile range. They are defined by the equations:

Sample variance for variable *X*:

$$s_x^2 = \frac{\sum_{i=1}^{n} (x_i - \bar{x}_i)^2}{\mathrm{n} - 1}$$

Sample standard deviation for variable *X*:

$$s_x = \sqrt{s_x^2}$$

While the mean shows the center of gravity in a variable, variance and standard deviation show the spread around the mean. Imagine that we test the quality of two kinds of tea with 100 hotel guests. According to Table 8.4, they have identical means of 5.6. Our immediate conclusion could be that the hotel guests agree that the teas are the same quality since their means are identical.

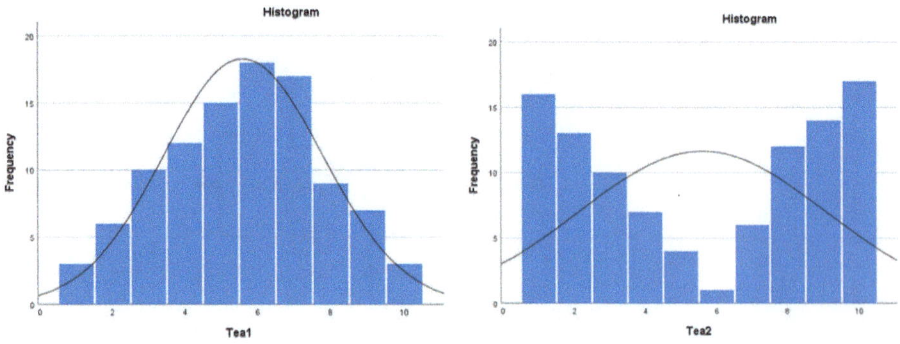

Fig. 8.13 Mean and distribution comparing two teas

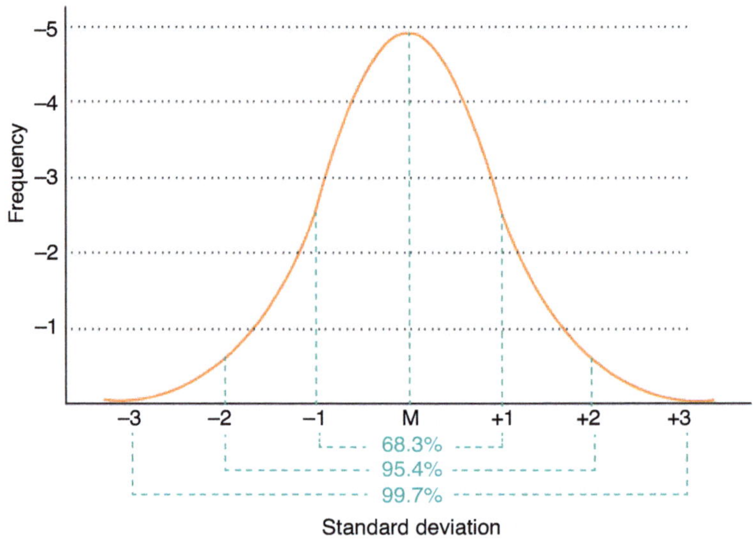

Fig. 8.14 Normal distribution and standard deviation

Figure 8.13 shows the histograms for the hotels guest's opinions of the two teas. Clearly, the distributions are extremely different. The standard deviation for Tea 2 is substantially higher, which is reflected in the data grouping to either end of the histogram on the right. The interpretation would be that for Tea 1, the guests perceive the quality in a uniform way. For Tea 2, several guests loved it and several guests hated it. If you have ever tried Rooibos tea, perhaps you will understand. It is an acquired taste that many people enjoy. For many, however, it tastes awful.

Another thing to learn from Fig. 8.13 is to not blindly trust the software. We requested a normal distribution curve to be imposed on both histograms. For Tea

2, the normal curve is meaningless, and even misleading. It does not at all reflect the true distribution of the data.

Standard deviation is the most common distribution measure, and expresses how much on average the observations for a variable deviate from the mean. In other words, it measures how closely the observations are gathered about the mean. Figure 8.14 shows how standard deviation is related to the normal distribution.

Assuming a variable with perfect normality, the graph shows the percentage of observations that will fall within one, two, and three standard deviations. The mean value (M) +/− one standard deviation will cover 68.3% of all observations. Two standard deviations from the mean will cover 95.4% of the observations, and three standard deviations will cover 99.7% of all observations. The size of the standard deviation depends on how the data is distributed (the variance) and on which scale we have measured the variable. We can thus use the mean and standard deviation of one variable to compare with another variable. From the tea example, Tea 1 has a mean of 5.6, +/− one standard deviation of 2.18, meaning that 68.3% of the observations will fall between 5.6-2.18=3.42 and 5.6+2.18=7.78. Tea 2 has a mean of 5.6, +/− one standard deviation of 3.43, meaning that 68.3% of the observations will fall between 5.6-3.43=2.17 and 5.6+3.43=9.03.

Note that in the equation for variance, we follow the tradition in statistics where we use n − 1 in the denominator. This is because we consider the variance (s_x^2) and the standard deviation (s_x) as estimates of the respective population values for population variance (σ^2) and population standard deviation (σ).

Example 8.1 Descriptive Statistics

We use the *Hotel* data and list some key descriptive statistics for the variable, number of employees.

In Table 8.5, the Case Processing Summary shows that there are no missing data in the 100 observations. In the Descriptives box, the mean is 32.76 with a standard deviation of 30.498. This is relatively large, which makes us suspicious about the normality of the distribution of the data. The median is 20.50, which is very different from the mean, also indicating an abnormal distribution.

The minimum number of employees is 2, and the maximum is 146, giving a range of 144. Since the interquartile range is so small (27) relative to the range, the data has extreme outliers. Since it is relatively low compared to the high end of the range, the extreme outliers will be at the high end of the range.

The histogram in Fig. 8.15 and the boxplot in Fig. 8.16 confirm the skew in the data. The boxplot identifies the case numbers of the outliers. ◀

Table 8.5 Case processing summary and descriptives for number of employees

Case Processing Summary

	Cases					
	Valid		Missing		Total	
	N	Percent	N	Percent	N	Percent
No.employ	100	100,0%	0	0,0%	100	100,0%

Descriptives

			Statistic	Std. Error
No.employ	Mean		32,76	3,050
	95% Confidence Interval for Mean	Lower Bound	26,71	
		Upper Bound	38,81	
	5% Trimmed Mean		29,18	
	Median		22,50	
	Variance		930,124	
	Std. Deviation		30,498	
	Minimum		2	
	Maximum		146	
	Range		144	
	Interquartile Range		27	
	Skewness		1,873	,241
	Kurtosis		3,361	,478

8.6 Analytical Techniques for Relationships between Variables

Measuring Reliability

There are two basic methods for estimating how reliable an empirical measurement is. In one case, a) one examines the consistency of a measurement over time. In the second case, b) one examines the consistency of different measures that measure the same thing. For questionnaires, it is the internal consistency of different questions that are meant to measure the same thing at a given time. This is especially relevant when measuring attitudes using multiple questions.

(a) Consistency over Time

Two variants are usually distinguished when it comes to mapping consistency over time: the *test-retest method* and the *alternative form method.* The test-retest method is to conduct a new survey in which the same questions are asked to the same people shortly after the first survey was administered. We then calculate the correlation

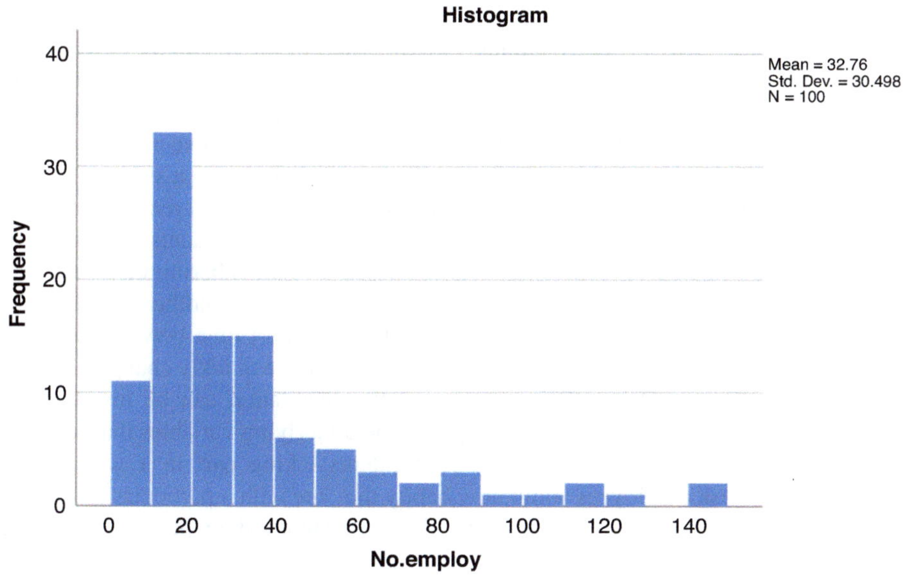

Fig. 8.15 Histogram for number of employees

Fig. 8.16 Boxplot for number of employees

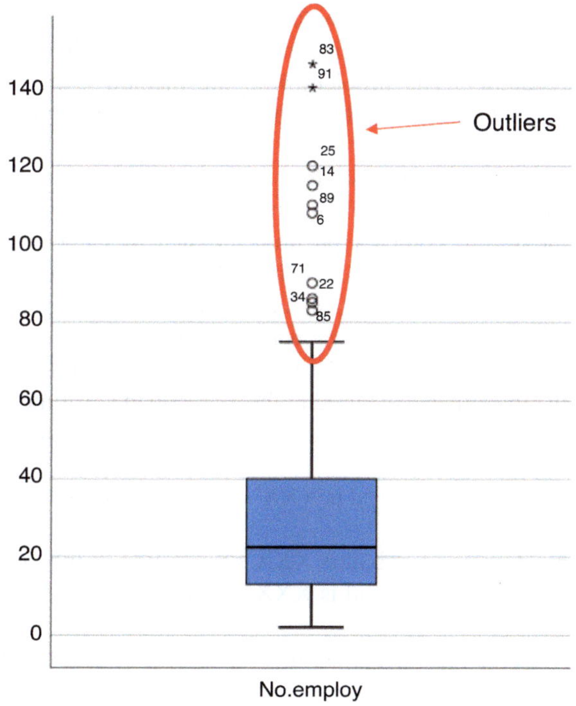

between the responses to the two questionnaires. The correlation coefficient, which can range from −1 to +1, is a quantitative expression of the reliability of the study (correlation is explained later in this chapter). With perfect reliability, the correlation coefficient will be +1. In this case, the results of the two studies may not be identical, but they will be consistent. In other words, if the relationship between the results of the two surveys is presented in a diagram, the answers will be on a straight line.

Although this method has an intuitive appeal, there are several reasons why it is not often used. It costs money and takes time. When respondents are asked to re-answer the same survey, they may answer based on remembering their previous answers instead of actually thinking about the questions. Another problem, of course, is that their opinion may have changed in the short time between surveys. For example, their attitude towards a job position may have quickly changed if other prospects became available. This means that the correlation can be low without being due to low reliability. In fact, changes in the underlying variables that we wish to identify may simply be due to the respondents taking part in a survey. For example, if we ask employees about how happy they are with a particular employer, this can lead to a thought process where they re-evaluate their perception of that employer.

The alternative form method differs from the test-retest method by using different questions each time the data is collected. The questions used the second time are assumed to cover the same theoretical properties or variables as the questions used the first time. The advantage of the alternative form method is foremost to reduce the risk that the respondent's memory will affect the correlation, and thus the estimate of reliability. The problem with the method is that it can be difficult to construct different questions that express the same underlying variable.

(b) Internal Consistency

In Chap. 6, we explained how Likert scales are used to measure constructs. When measuring complex phenomena such as values, lifestyles, and attitudes, using a single question is insufficient. As a rule of thumb, multiple measures (sometimes called indicators) are used to capture various aspects of the theoretical concepts we wish to measure. By evaluating the internal consistency of the measures of a construct, we assess the reliability of the overall measurement. As an example, we list three questions used in the Norwegian Customer Barometer index (NKB), which together measure corporate image (reputation) in the residential alarm market.

- How good or bad is XXX's reputation compared to other Norwegian home alarm suppliers?
- How good or bad is XXX's reputation in the opinion of your friends?
- How good or bad is XXX's reputation in general?

All the questions are measured on a 10-point scale with anchors where 1 is very poor and 10 is very good.

Split-halves reliability is a simple method to measure the internal consistency of a set of measures, in this case, questions. Start by randomly dividing the questions

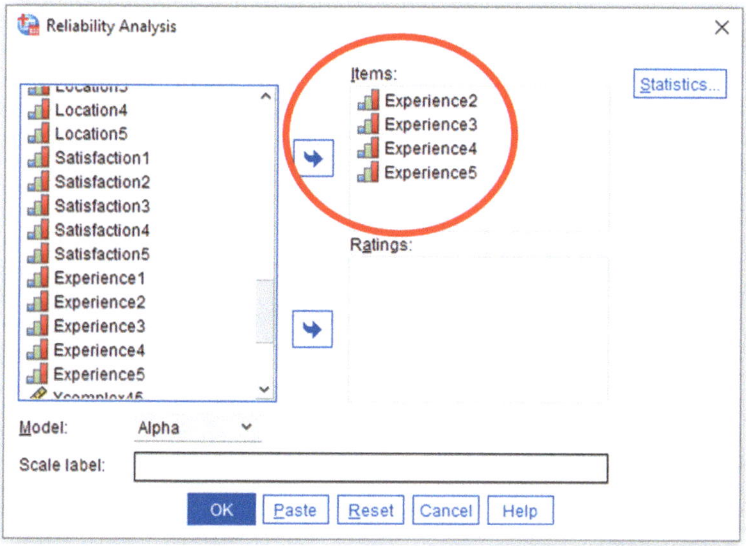

Fig. 8.17 Reliability analysis

about a specific construct into two groups. Collect data, and then calculate the average scores for each group of questions. Calculate a correlation coefficient for the averages for the two groups. A high correlation indicates high internal consistency in the overall scale, and thus, reliability. If the correlation is low, for example, below an absolute value of 0.5, then something is wrong with one or more of the questions. For the corporate image questions, we would collect data, randomly choose two and calculate their average, and then calculate the correlation between the average and the other question. Note, the split-halves method is more suitable when a large number of questions are used to measure a phenomenon.

The problem with the split-halves technique is that, depending on the number of measures, there are many different ways that they can be divided into two groups. With our simple example with just three questions, there are three possible groupings (1 to 2 + 3; 2 to 1 + 3; 3 to 1 + 2). The reliability measure will vary depending on how the questions are grouped. To address this, techniques have been developed to measure reliability based on the average correlation between all measures of a construct. The best known is *Cronbach's alpha*. In addition to calculating the reliability, the reliability if a specific measure is deleted can also be calculated. This allows the researcher to evaluate whether to drop an indicator in order to increase the reliability of the overall measure. Note: you can only calculate Cronbach's alpha for the indicators of a single construct (i.e., one construct at a time). A more detailed discussion is given by Carmines and Zeller (1979).

(c) Cronbach's Alpha

Cronbach's alpha is the most widely used measure of reliability, often referred to as simply, alpha (α), It can be calculated according to the formula where *a* is the number of indicators and *b* is the sum of the correlations between the indicators:

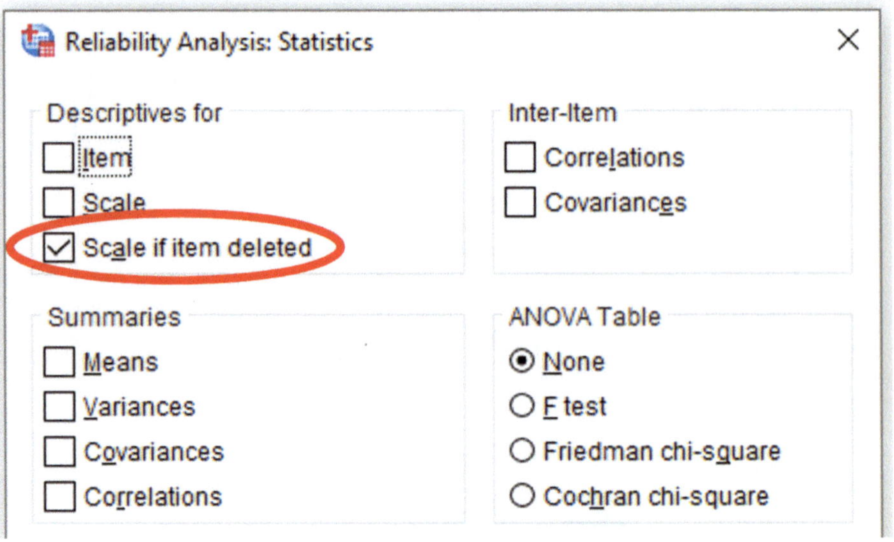

Fig. 8.18 Reliability analysis: scale if item deleted

Table 8.6 Reliability output: Cronbach's alpha

Reliability Statistics

Cronbach's Alpha	N of Items
.799	4

Item-Total Statistics

	Scale Mean if Item Deleted	Scale Variance if Item Deleted	Corrected Item-Total Correlation	Cronbach's Alpha if Item Deleted
Experience2	15.99	9.927	.594	.758
Experience3	15.85	9.007	.745	.684
Experience4	15.91	8.252	.693	.708
Experience5	15.97	11.259	.439	.825

Table 8.7 Reliability output: Cronbach's alpha

Reliability statistics	
Cronbach's Alpha	No of items
0.825	3

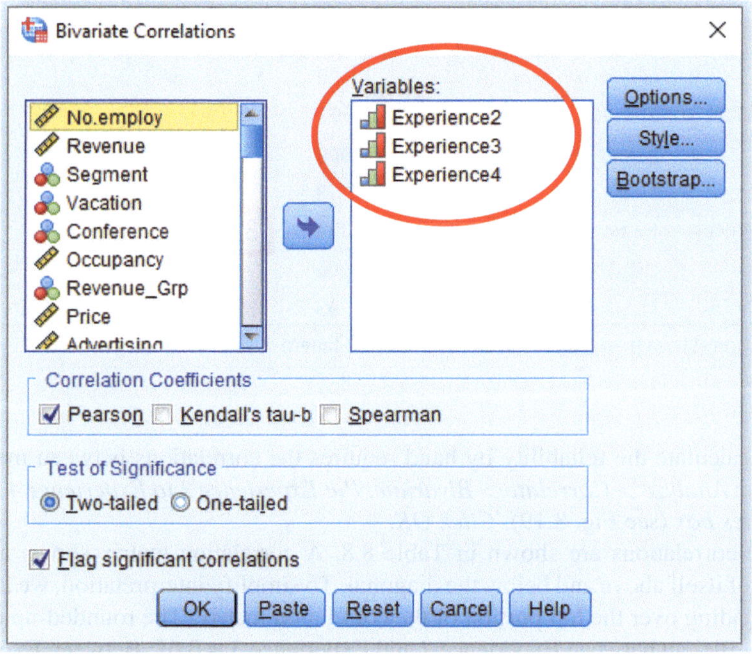

Fig. 8.19 Estimating correlations in SPSS

$$\alpha = \frac{a}{a-1}\left(1 - \frac{a}{a+2b}\right)$$

First, we will show how to estimate Cronbach's alpha in SPSS; then we will go through the calculations based on the formula and correlations. In the *Hotel* data, choose: *Analyze > Scale > Reliability Analysis. Put Experience 2 to Experience 5 into the items box* (see Fig. 8.17). *Choose statistics > tick Scale if item deleted > Continue > OK* (see Fig. 8.18).

The results in Table 8.6 show that the Cronbach's alpha measure of reliability is 0.799, which is well above the threshold recommended by Nunnally (1978). However, at the bottom right the table shows that the reliability of the scale will improve to 0.825 by dropping Experience 5. The question for the researcher is whether it is better to have higher reliability with just three questions. If the remaining questions, Experience 2 to Experience 4, sufficiently represent the dimensions of the underlying construct, then it is OK to drop Experience 5.

Table 8.7 shows the Cronbach's alpha (0.825) with only Experience 2 to Experience 4.

Table 8.8 Correlations

Correlations

		Experience2	Experience3	Experience4
Experience2	Pearson Correlation	1	.666**	.539**
	Sig. (2-tailed)		.000	.000
	N	99	99	99
Experience3	Pearson Correlation	1 .666**	1	.632**
	Sig. (2-tailed)	.000		.000
	N	99	100	100
Experience4	Pearson Correlation	2 .539**	3 .632**	1
	Sig. (2-tailed)	.000	.000	
	N	99	100	100

**. Correlation is significant at the 0.01 level (2-tailed).

To calculate the reliability by hand requires the correlations between measures. Choose: *Analyze > Correlate > Bivariate. Put Experience 2 to Experience 4 into the variables box* (see Fig. 8.19). *Click OK.*

The correlations are shown in Table 8.8. A correlation matrix shows a mirror image of itself above and below the diagonal. To simplify interpretation, we have put blue shading over the top portion of the correlation matrix. The rounded-up correlation coefficient between Experience 2 and Experience 3 is 0.67. Between Experience 2 and Experience 4 it is 0.54, and between Experience 3 and Experience 4 it is 0.63. The double asterisk means that the correlation is significant at the 0.01 level. The significance is also shown as sig., and we normally look for values below 0.05. More specifically, this means that at a specified significance level we reject the null hypothesis that the correlation is equal to zero. N indicates the number of observations in the variable.

To calculate Cronbach's alpha we put the information into the formula where:

$a = 3$,
$b = (0.67 + 0.54 + 0.63) = 1.84$.

In the formula:

$$\alpha = \frac{a}{a-1}\left(1 - \frac{a}{a+2b}\right) = \frac{3}{3-1}\left(1 - \frac{3}{3+2*1.84}\right) = 0.83$$

Referring back to Table 8.7, the Cronbach's alpha is 0.825, which when rounded up equals our hand-calculated version, 0.83.

Chronbach's alpha is a function of the number of indicators and the correlation between them. As the number of indicators increases and the correlations increase,

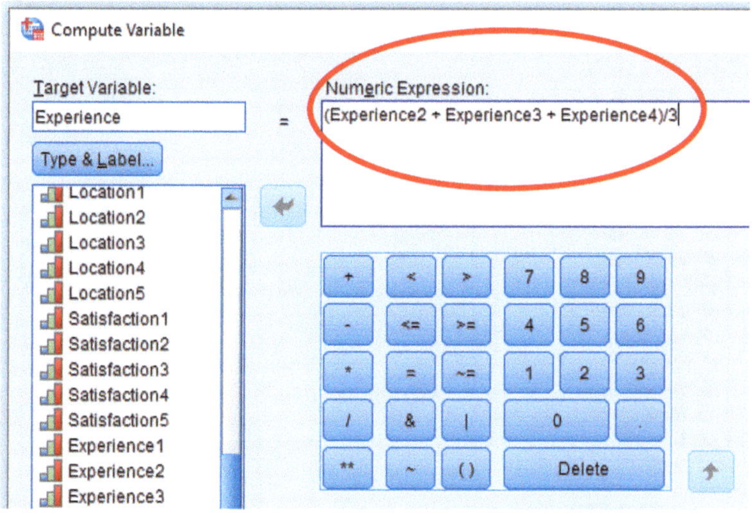

Fig. 8.20 Compute variable

alpha will increase. If the correlations between indicators are too high, they are too similar to capture unique dimensions of the construct. This problem increases as the number of indicators increases. There is no absolute value; however, alpha should not be too close to 1. Remember! Cronbach's alpha can only be estimated for one set of measures at a time. It is a typical beginner's mistake to combine measures for constructs and then estimate alpha.

Creating New Variables

When several indicators reliably measure a construct, it is common to add them together for further analysis. In Fig. 8.20, we show how to compute several indicators into a single variable. Choose: *Transform > Compute Variable.*

The next step is to (see Fig. 8.20):

1. Put a name in the Target Variable box. We chose Experience.
2. Move the indicators (Experience1, Experience2, and Experience3) into the Numeric Expression box, while at the same time,
3. Add the proper numeric expressions.

By adding the three questions together inside brackets and then dividing by the number of questions (3), the range of the new variable is the same as the source variables. The new *Experience* variable can be used to represent the three underlying indicators in further analyses.

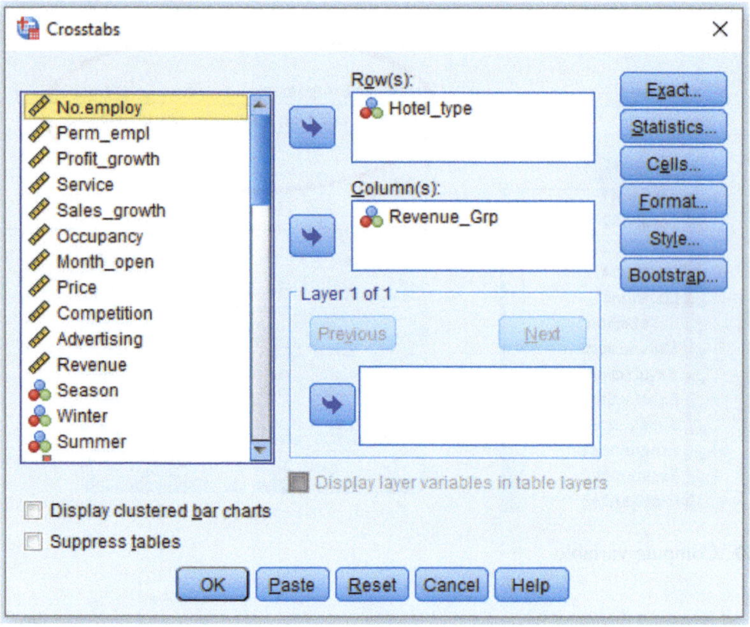

Fig. 8.21 Cross-tabulation

Table 8.9 Cross-tabulation output

Hotel_type * Revenue_Grp Crosstabulation				
		Revenue_Grp		
Count		Low	High	Total
Hotel_type	Boutique	34	6	40
	Chain	32	28	60
Total		66	34	100

Cross-Tabulation

Cross-tabulation tables are appropriate for studying the relationship between nominal level (categorical) variables. They provide an overview as well as tests for statistically significant patterns. Open the *Hotel* data. The *Revenue* variable refers to the amount of money that each hotel has taken in during one year. The variable, *Revenue_Grp*, is derived from the Revenue variable, by dividing it into two categories: low revenue (0–20 million) and high revenue (31 million and over). There are 66 observations in the low revenue group and 34 in the high revenue group.

In the cross-tabulation, we want to investigate the connection between revenue groups and hotel types. There are 40 boutique hotels and 60 branded chain hotels (variable Hotel_type). Choose: *Analyze > Descriptive Statistics > Crosstabs > move Hotel_type to Row(s) and Revenue_Grp to Column(s)* (see Fig. 8.21).

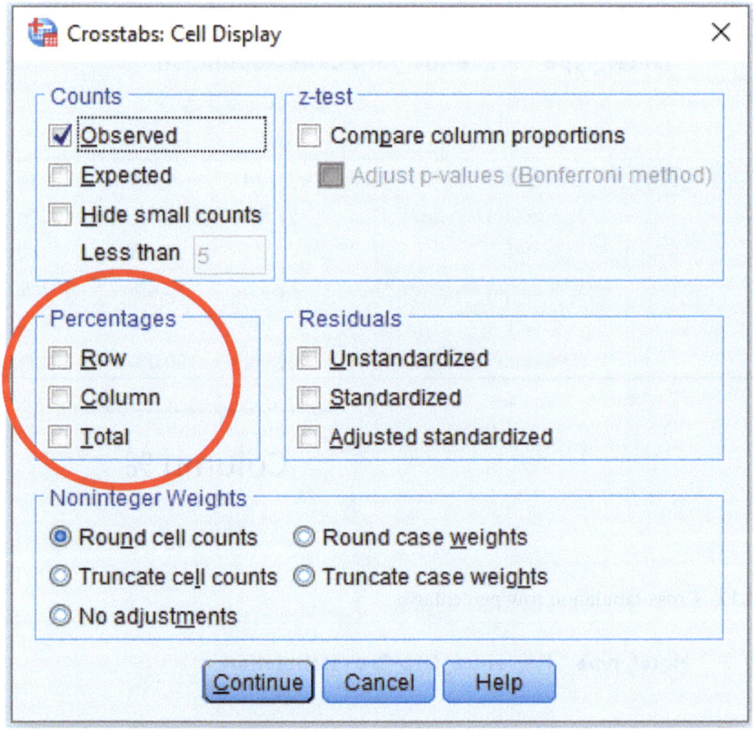

Fig. 8.22 Cross-tabulation percentages

The cross-tabulation output in Table 8.9 shows 34 boutique hotels and 32 chain hotels in the low revenue group. In the high revenue group, there are only 6 boutique hotels compared with 28 chain hotels. There appears to be a pattern where boutique hotels are more frequently associated with low revenue and chain hotels are distributed equally between low and high revenue groups. In Chap. 9, we will return to this example and show how with the chi-square statistic we can test whether the differences are statistically significant. Note that with this example, which applies to cross-tabulation in general, we do not infer any causal relationships. As with correlation, *we test association, not causality*. As an example, imagine that it happens to be more likely that large hotels choose to belong to chains, while small hotels choose to remain independent. This is called *self-selection bias*. It happens when the units of analysis in some way self-select which group they belong to.

When running cross-tabulations, the interpretation greatly depends on how the table is presented. It can help to present totals by percentage, either in columns or in rows. In Cross-tabs, choose: *Cells > and then tick rows, columns, or totals* (see Fig. 8.22).

Note that we said "or"! Always remember that you are trying to convey the meaning in the data. More is not necessarily better. In the current example, it may be

Table 8.10 Cross-tabulation column percentages

Hotel_type * Revenue_Grp Crosstabulation

			Revenue_Grp		
			Low	High	Total
Hotel_type	Boutique	Count	34	6	40
		% within Revenue_Grp	51.5%	17.6%	40.0%
	Chain	Count	32	28	60
		% within Revenue_Grp	48.5%	82.4%	60.0%
Total		Count	66	34	100
		% within Revenue_Grp	100.0%	100.0%	100.0%

Column %

Table 8.11 Cross-tabulation row percentages

Hotel_type * Revenue_Grp Crosstabulation

			Revenue_Grp		
			Low	High	Total
Hotel_type	Boutique	Count	34	6	40
		% within Hotel_type	85.0%	15.0%	100.0%
	Chain	Count	32	28	60
		% within Hotel_type	53.3%	46.7%	100.0%
Total		Count	66	34	100
		% within Hotel_type	66.0%	34.0%	100.0%

Row %

best to show percentages in columns (see Table 8.10), or in rows (see Table 8.11). It depends on what you are trying to convey.

In Table 8.10, the revenue groups total to 100% in the columns. The interpretation is that in the low revenue group, the hotel types are fairly equally divided (51.5% to 48.5%). In the high revenue group, only 17.6% are boutique hotels while 82.4% are chain hotels. Clearly, there is an imbalance in the high revenue group.

In Table 8.11, the hotel types total to 100% in the rows. For boutique hotels, 85% are low revenue and 15% are high revenue. For chain hotels, slightly more are in the low revenue groups (53.3%) than the high revenue group (46.7%).

The choice of interpreting rows or columns is up to which nuance the researcher wants to express with the data.

Table 8.12 Calculating correlation

	X	$X-\overline{X}$	$(X-\overline{X})^2$	Y	$Y-\overline{Y}$	$(Y-\overline{Y})^2$	$X * Y$
	23	−10.2	103.4	15	−8	62.7	80.5
	30	−3.2	10.0	25	2	4.3	−6.6
	34	0.8	0.7	20	−3	8.5	−2.4
	20	−13.2	173.4	10	−13	166.8	170.1
	30	−3.2	10.0	19	−4	15.3	12.4
	110	76.8	5903.4	75	52	2712.7	4001.7
	6	−27.2	738.0	4	−19	357.8	513.9
	27	−6.2	38.0	18	−5	24.2	30.3
	14	−19.2	367.4	8	−15	222.5	285.9
	19	−14.2	200.7	24	1	1.2	−15.3
	60	26.8	720.0	41	18	327.0	485.2
	25	−8.2	66.7	16	−7	47.8	56.5
Mean	33.2		8331.7	23		3950.9	5612.2

$Corr(X,Y) = \frac{5612.2}{\sqrt{8331.7 * 3950.9}} = 0.978$

Covariance and Correlation

Covariance and correlation as analysis techniques are experiencing immense growth in connection with analyzing big data. Often, they are used to explore for patterns in data, and many analyses are no more advanced than what we describe in this chapter. In some circumstances, clustering and factor analysis are also used (both are described later in the book). It is important to be aware that despite the large datasets associated with big data, many of the basic statistics are the same. Understanding the difference between covariation and causality between variables is even more important than before.

Covariance and *correlation* both measure the linear relationship between variables. The Pearson correlation (developed by Karl Pearson) is a standardized version covariance, with its values restricted within the range of −1 to +1. When people talk about correlation, they are most often referring to the Pearson correlation. While both covariance and correlation indicate the positive or negative direction of the relationship between two variables, correlations also provide a measure of the strength of the relationship. The covariance between two variables X and Y for a sample is defined by:

$$Cov\,(X,Y) = \frac{\sum_{i=1}^{n}\left(X-\overline{X}\right)\left(Y-\overline{Y}\right)}{n-1}$$

The correlation is defined by:

Table 8.13 Pearson correlation matrix

Correlations

		No.employ	Revenue
No.employ	Pearson Correlation	1	.924**
	Sig. (2-tailed)		.000
	N	100	100
Revenue	Pearson Correlation	.924**	1
	Sig. (2-tailed)	.000	
	N	100	100

**. Correlation is significant at the 0.01 level (2-tailed).

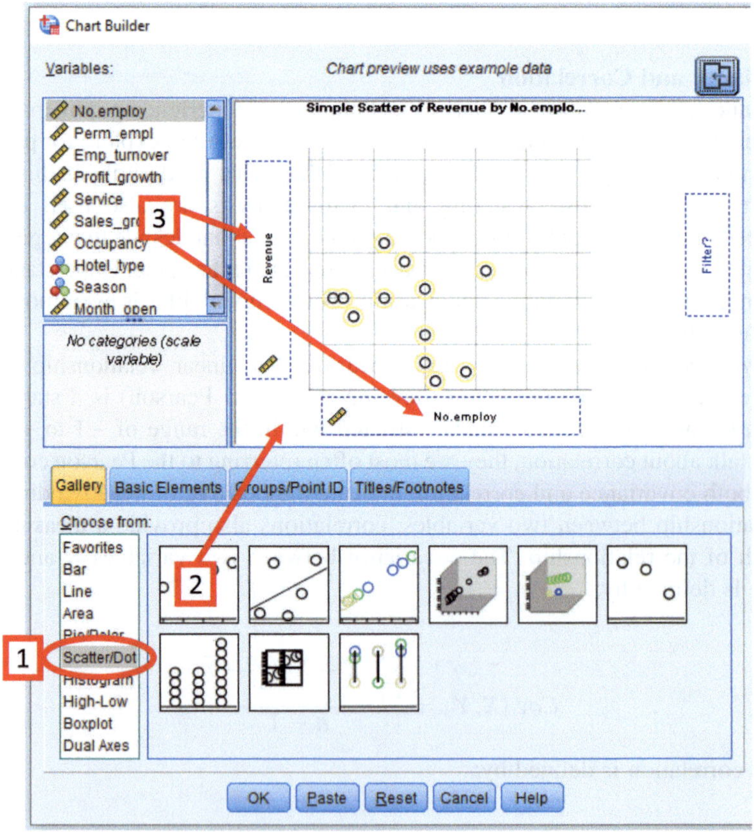

Fig. 8.23 How to estimate a correlation scatterplot

Fig. 8.24 Scatterplot of revenue and number of employees

Table 8.14 Correlation strength

Correlation strength	Absolute value of correlation coefficient
Small	0.10–0.29
Medium	0.30–0.49
Large	0.50–1.00

$$Corr\ (X, Y) = \frac{Cov(X, Y)}{s_X s_Y} = \frac{\sum_{i=1}^{n} (X - \overline{X})(Y - \overline{Y})}{\sqrt{\sum_{i=1}^{n} (X - \overline{X})^2 \sum_{i=1}^{n} (Y - \overline{Y})^2}}$$

Pearson Correlation

In this example, we consider the association between the number of employees at a hotel (X) and the revenue (Y). It makes sense that higher revenue comes from larger hotels with many guests, and therefore, many employees. We show how to calculate the correlation based on the first 12 observations in the *Hotel* data. Then, we compare it with the correlation estimated in SPSS from all 100 observations in the dataset. By breaking down the formula into parts, it is easier to see patterns. The calculations are as follows in Table 8.12:

In SPSS using the *Hotel* data, choose: *Analyze > Correlate > Bivariate > move No.employ and Revenue to the Variables box and click OK.* The correlation from

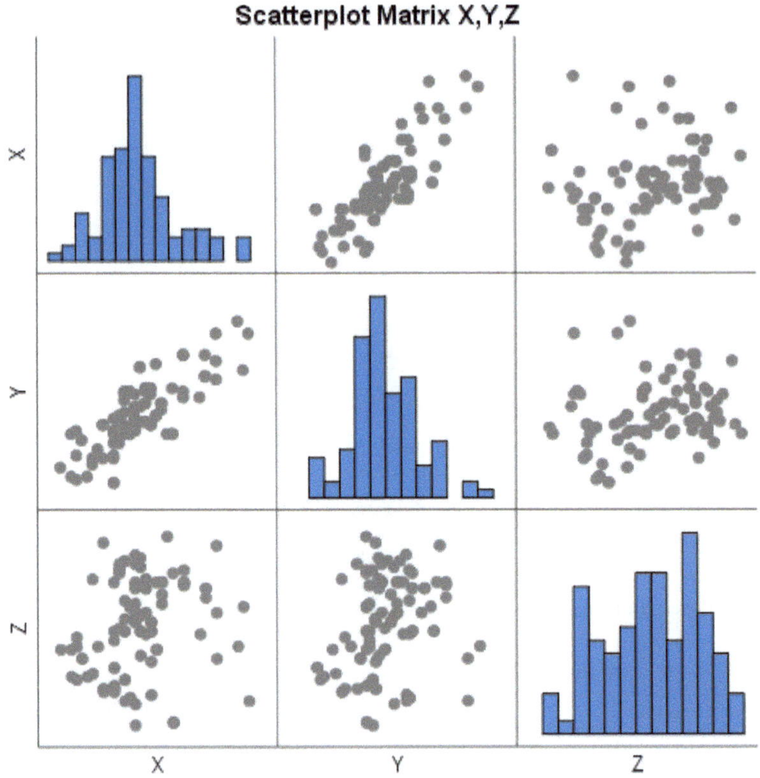

Fig. 8.25 Scatterplot matrix

SPSS using all 100 observations is 0.924 (see Table 8.13), which is reasonably similar to the correlation we calculated using the first 12 observations (0.978).

To make a scatterplot of the correlation, choose: *Graphs > Chart Builder > then choose (1) Scatter/Dot, then (2) drag the simple scatterplot up to the open box, then (3) drag the respective variable to the axis boxes* (see Fig. 8.23).

Figure 8.24 shows the scatterplot for the number of employees and revenue at hotels in the sample. Not surprisingly, given the high correlation coefficient, it shows a very distinct positive linear relationship. The *Fit Line* gives an idea of the linear relationship between the variables (positive slopes up, negative slopes down). The y = 0.86 indicates where the Fit Line intersects the Y-axis, and the 0.61*x indicates the slope of the line.

The correlation between variables can be assessed for statistical significance (significance is discussed in the next chapter). However, correlation coefficients should also be evaluated for their practical significance as well. Cohen (1988) established practical guidelines for interpreting statistically significant correlations. As a rule of thumb, correlations below 0.1 are meaningless. From 0.1 to 0.29, they are small. From 0.3 to 0.49, they are moderate, and 0.5 and over are large (see

Table 8.15 Observations for Spearman's rank correlation

Observation	*X*	*Y*
1	2	3
2	4	5
3	5	4
4	1	3
5	6	6
6	7	5
7	2	1
8	3	2

Table 8.16 Ranked observations

Rank *X*	Rank *Y*	d	d^2
2.5	3.5	−1.0	1.00
5.0	6.5	−1.5	2.25
6.0	5.0	1.0	1.00
1.0	3.5	−2.5	6.25
7.0	8.0	−1.0	1.00
8.0	6.5	1.5	2.25
2.5	1.0	1.5	2.25
4.0	2.0	2.0	4.00
Total $\sum$			20.00

Table 8.14). The correlation coefficient from the full dataset (0.924) is clearly in the large category.

A *scatterplot matrix* can be a good way to simultaneously show the characteristics of several variables. The plot in Fig. 8.25 is a scatterplot matrix for three random variables *X*, *Y*, and *Z* (data not supplied). The histograms on the diagonal provide information on the normality of the distribution for each variable. On the off-diagonal, the relationship between *Y* and *X* appears to be quite strong (correlation = 0.794), whereas the relationships between *X* and *Z*, and *Y* and *Z*, appear much weaker (correlation *X* and *Z* = 0.226, correlation *Y* and *Z* = 0.258).

Spearman's Rank Correlation

The *Spearman's rank correlation* is for estimating a correlation coefficient for ordinal level variables. The Pearson correlation coefficient assumes data at the interval or ratio level. Often, when variables are measured on Likert-type scales, they are treated as *approximately interval* and Pearson correlation coefficients are estimated. To a great degree, this is justified when there are no comparable nonparametric techniques. In this case, however, the Spearman rank correlation is an excellent alternative, and in fact, the correct choice for all ordinal variables including Likert scales. It is defined by the equation:

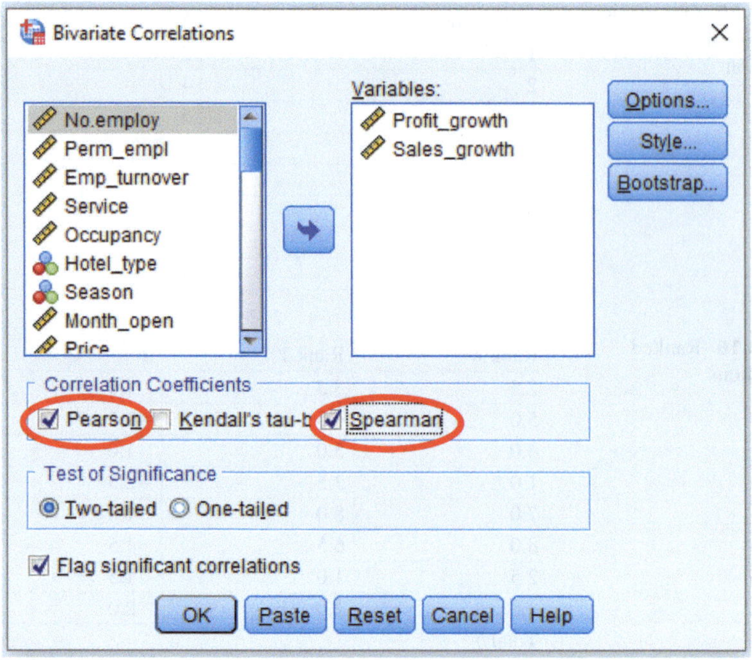

Fig. 8.26 Spearman's rank correlation in SPSS

$$SR = 1 - \frac{6 \sum d_i^2}{n(n^2 - 1)}$$

Where d_i = the difference between the ranks of corresponding variables and n is the sample size. To calculate the rank correlation between two variables *X* and *Y*, start by ranking the observations. This means assigning the highest ranking to the highest *X*-value and the highest *Y*-value. Then the second highest *X*- and *Y*-values get the second highest ranking, and so on. Continue until all *X*- and *Y*-values have been assigned a rank. An example will help explain this.

There are 8 observations (see Table 8.15). The highest number in each variable *X* and *Y* gets assigned the rank number 8. The next highest numbers are assigned 7, and so on until rank 1.

The highest *X*-value is 7 (observation #6), so it is assigned the rank of 8. The highest *Y*-value is 6 (observation #5), so it is also assigned the rank of 8. When variable values are equal, they receive the same rank; however, the sum of the ranks remains the same.

$$1 + 2 + 3 + \ldots + n = \frac{n(n+1)}{2}$$

Table 8.17 Pearson correlation output

Correlations

		Profit_growth	Sales_growth
Profit_growth	Pearson Correlation	1	,432**
	Sig. (2-tailed)		,000
	N	100	100
Sales_growth	Pearson Correlation	,432**	1
	Sig. (2-tailed)	,000	
	N	100	100

**. Correlation is significant at the 0.01 level (2-tailed).

Table 8.18 Spearman's correlation output

Correlations

			Profit_growth	Sales_growth
Spearman's rho	Profit_growth	Correlation Coefficient	1,000	,435**
		Sig. (2-tailed)	.	,000
		N	100	100
	Sales_growth	Correlation Coefficient	,435**	1,000
		Sig. (2-tailed)	,000	.
		N	100	100

**. Correlation is significant at the 0.01 level (2-tailed).

In Table 8.15, the X variable has the value 2, twice. Both are assigned the value 2.5, such that 2.5 + 2.5 = 5 (see Table 8.16). Had there been minute differences between the two values, such as 1.99 and 2.01, the highest value would be assigned rank 3 and the slightly lower value assigned rank 2, such that the sum of the numbers would equal 5.

$$SR = 1 - \frac{6 * \sum_{i=1}^{n} d_i^2}{n^3 - n} = 0.758$$

The Pearson correlation for the same data is 0.769, which is not far from the Spearman correlation of 0.758.

Example 8.2

We estimate the Spearman's rank correlation with the *Hotel* data for the variables Profit_growth and Sales_growth. Choose: *Analyze > Correlations > Bivariate > and tick the box for Spearman* (see Fig. 8.26).

The output in Table 8.17 shows the Pearson correlation coefficient (0.432), and the output in Table 8.18 shows the Spearman's correlation coefficient (0.435).

Just like with a Pearson correlation, the Spearman correlation ranges from −1 to +1. Zero means no correlation, and an absolute value of 1 means perfect correlation. A value of 0.435 indicates a medium strength relationship between the two variables. ◀

8.7 Summary

The purpose of this chapter was to show simple data analysis techniques for getting to know a dataset. Starting with cleaning the data, the researcher, as much as possible, wants to remove errors from the dataset. Graphical representation is a good tool for both cleaning and getting to know the data.

For analyzing single variables, we looked at frequency distributions and bar charts for two reasons. First, they are simple to create and understand, and second, they provide information on the distributional characteristics of the variables. There is vast potential for graphical representation, far greater than what we presented here. We looked at the most common measures of central tendency: the mean and the median. For measures of spread, we looked at the variance, standard deviation, interquartile range, and range.

For analyzing the relationship between two variables, we started with considering reliability and Cronbach's alpha. Often, several variables indicate a higher-level variable. Reliability provides a measure of the internal consistency of the variables that are indicators of the same construct. Cross-tabulation tables are a way to analyze the relationship between categorical variables, most often nominal variables. For ordinal variables, Spearman's rank correlation is appropriate. For continuous variables, we discussed Person correlations. Both Pearson and Spearman correlations share the characteristic that they range from −1 to +1, where zero means no relationship. −1 means a perfect negative relationship, while +1 means a perfect positive relationship.

References

Aguinis, H., Gottfredson, R. K. & Joo, H. (2013). Best-Practice Recommendations for Defining, Identifying, and Handling Outliers. *Organizational Research Methods*, 16(2), 270-301.

Carmines, E. G., & Zeller, R. A. (1979). *Reliability and validity assessment*. Sage.

Cohen, J. (1988). *Statistical power analysis for the behavioral sciences*. L. Erlbaum Associates.

Nunnally, J. C. (1978). *Psychometric theory*. McGraw-Hill.

Hypothesis Testing

9

Contents

9.1 Introduction

If you approach your career as a business model, that is, how you create value to make money, then consider the vital link between analytical expertise and business knowledge. Unlocking the value in big data requires analytical skills coupled with an understanding of how to apply it in business. Data alone does not create value. You, along with a computer, process and interpret data to provide valuable insights. Negligent misuse of data often causes more harm than good.

This chapter is presented in two parts. First, you will learn how to test a hypothesis on a representative sample of a population, and then how to test hypotheses about relationships between groups and variables (see Fig. 9.1). We use SPSS in the examples.

Often, data for the variables we measure comes from a sample (a small part) of the population. We discussed populations and samples in Chap. 7. When data is from a sample, we usually want to test whether we can generalize conclusions from what we find in the sample to the entire population. In this chapter, we will learn how through hypothesis testing it is possible to use statistics to determine whether the relationships we observe are indeed statistically significant for the population. If we find, for example, that hotel room prices are lower in areas where competition is high, we want to know how confident we can be that this exists throughout the entire

J. E. Sallis et al., *Research Methods and Data Analysis for Business Decisions*, Classroom Companion: Business, https://doi.org/10.1007/978-3-030-84421-9_9

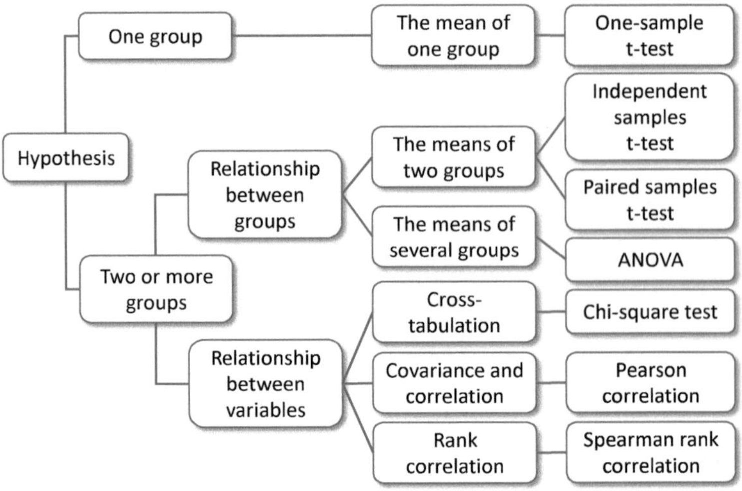

Fig. 9.1 Techniques for hypothesis testing

region (population). *Statistical inference* is concerned with drawing conclusions about one or more characteristics of the population on the basis of representative samples from one or more populations (it is often appropriate to consider the population as infinitely large).

In many cases, one will test whether there is a difference between two means or test the level of correlation between two variables. These characteristics of a population are often called *parameters*. *Hypothesis testing* is a procedure to test whether a population parameter is different from a hypothesized value. Hypothesis testing is a way to evaluate theory. We follow the tradition in social sciences where two hypotheses are formulated: the H_0 *null hypothesis* and the H_1 *alternative hypothesis*. The alternative hypothesis represents the theory we want to test, and the null hypothesis is that there is no significant difference between the parameters we are testing. In other words, the null hypothesis rejects our theory. The only exception to this rule is in structural equation modeling (SEM) where the rule is reversed.

The null hypothesis typically specifies the range of values that the parameter can be expected to assume if the theory does not hold, for example, that there is no difference between groups or no correlation between the variables we test. The alternative hypothesis specifies the range of values that are consistent with the theory, for example, that there is a difference between groups or the variables are correlated. A *theory* is an argument or rationale that explains something. *Hypotheses* are testable statements for evaluating a theory.

9.2 Hypothesis Tests and Error

We use the following notation:

Corr (X, Y)	Correlation between X variable and Y variable
H_0	Null hypothesis
H_1	Alternative hypothesis (also denoted HA)
$\|t\|$	Test-statistic (absolute value) for the *t* distribution
t_α	Critical value for the *t* distribution
α	Alpha = significance level = the probability of type I error
β	Beta = the probability of type II error
$1-\beta$	Power

In Chap. 8, as an example of correlation we proposed a relationship between the number of employees and the revenue level in a sample of hotels. Our argument was that higher revenue is associated with larger hotels that have many employees. This suggests a positive correlation, meaning that the two variables rise together and fall together. This can be tested by measuring the correlation between the variables. If the correlation coefficient is statistically significant (i.e., significantly greater than zero), then the null hypothesis is rejected, and the alternative hypothesis that there is a positive correlation between the number of employees and revenue is supported. The formal hypothesis setup looks like this:

H_0: *Corr (No.employ, Revenue)* = 0
H_1: *Corr (No.employ, Revenue)* > 0

The null hypothesis (H_0) states that the correlation between the number of employees (No.employ) and revenue (Revenue) is equal to zero. In other words, there is no correlation. The alternative hypothesis (H_1) states that the correlation between the number of employees (No.employ) and revenue (Revenue) is greater than zero. In other words, there is a significant correlation. Only one of the hypotheses can be supported. The test itself is performed by calculating a *test-statistic,* $|t|$, from the sample. The test-statistic has a probability distribution based on the null hypothesis being true. *Note that we assume that the null hypothesis is true.*

Theoretical Significance and Statistical Significance
In studies that intend to generate new theoretical knowledge, it is particularly important to be aware of the difference between theoretical and statistical significance. Even though a study might show that some hypotheses are statistically significant, it does not mean that there is theoretical evidence to support the findings. As our discussion of sample size and statistical power indicated, if the sample size is large enough, the alternative hypothesis, which is usually what we are looking for, will be supported. In this new age of big data, we can find statistical support for almost anything we want, even when it is meaningless. Researchers, especially when doing statistical analyses on large datasets, must always take a critical view of their findings and decide to what extent the findings are meaningful. A common mistake is to not test for spurious effects. This means incorrectly attributing outcomes to the wrong variables without testing for other relevant explanatory variables. We return to these issues when we discuss regression analysis in Chap. 10.

Understanding the Significance Level

How certain must we be to claim that there is a connection between events? In a significance test, we test whether we can reject the null hypothesis. If there is 100% certainty that the null hypothesis is true, the reported significance value will be 1.00. If there is 0% certainty that the null hypothesis is true, the reported significance value will be 0.00. However, we are never absolutely certain of the truth, so we never use these values. Instead, by taking one minus the reported significance level (e.g., $1 - 0.05 = 0.95$, or 95%), we get a measure of how confident we are that the alternative hypothesis is supported. The most commonly applied significance level in social sciences is 5%, meaning that we are 95% confident that the alternative hypothesis is true, and not just by chance.

Type I and Type II Errors

Imagine being a juror in a criminal trial. The consequence of the jury's decision has four possible outcomes. Two of the possibilities are in line with the truth: You can release an innocent person, or you can imprison a guilty person. However, imagine that the evidence in the case is not entirely clear. If you make the wrong interpretation, you run the risk of sentencing an innocent person to prison or releasing a guilty person. This illustrates Type I and Type II errors. *Type I error* is the most serious error and occurs when you reject a true H_0 (putting an innocent person in jail). *Type II error* occurs when you retain a false H_0 (freeing a guilty person).

In Chap. 8, as an example of correlation we proposed a relationship between the number of employees and the revenue level in a sample of hotels. Our argument was that higher revenue is associated with larger hotels that have many employees. This suggests a positive correlation, meaning that the two variables rise together and fall together. This can be tested by measuring the correlation between the variables. If the correlation coefficient is statistically significant (i.e., significantly greater than zero), then the null hypothesis is rejected, and the alternative hypothesis that there is a positive correlation between the number of employees and revenue is supported. The formal hypothesis setup looks like this:

H_0: *Corr (No.employ, Revenue)* $= 0$
H_1: *Corr (No.employ, Revenue)* > 0

The null hypothesis (H_0) states that the correlation between the number of employees (No.employ) and revenue (Revenue) is equal to zero. In other words, there is no correlation. The alternative hypothesis (H_1) states that the correlation between the number of employees (No.employ) and revenue (Revenue) is greater than zero. In other words, there is a significant correlation. Only one of the hypotheses can be supported. The test itself is performed by calculating a *test-statistic*, $|t|$, from the sample. The test-statistic has a probability distribution based on the null hypothesis being true. *Note that we assume that the null hypothesis is true.*

Table 9.1 Type I and type II errors

	H_0 is true	H_0 is false
Retain H_0	Correct	Type II error
Reject H_0	Type I error	Correct

Type I Error: Rejecting a True Null Hypothesis (H_0)
Whether the null hypothesis should be rejected or not depends on the size of the test-statistic, its probability distribution, and the selected significance level (α level), compared with the critical cutoff value. If the test-statistic is higher than the critical cutoff value, the null hypothesis is rejected, and we can state that H_1 is supported. In our correlation example, we would claim that there is a positive correlation between the number of employees and revenue in the hotel sample. If this is wrong, meaning that we claim that there is a correlation when in fact there is not, we have committed type I error.

Type II Error: Retaining a False (H_0)
If we instead retain H_0 when in fact H_1 is correct, we make a *Type II error*. In other words, a Type II error means that we retain a null hypothesis that is false. In our correlation example, this means we would claim that there is no correlation between the number of employees and revenue, when in fact there is.

Table 9.1 shows that there are four possible outcomes when testing a hypothesis. Type I and Type II errors represent two of the possibilities. If H_0 is true and the statistical test also indicates that H_0 should be supported, then a correct decision is made. A correct decision is also made if H_0 is false and the statistical test rejects H_0. It is important to understand that in hypothesis testing, there is a trade-off between the two types of error. Neither one can be completely eliminated.

9.3 The *T*-test

T-tests determine whether there is a statistically significant difference between the means of two groups or between the mean of one group and a specified test value. Sometimes they are referred to as the *Student's t*-test, which is the pseudonym for the person who developed them. Below, we explain the three main types of *t*-tests.

***T*-test for One Sample**
The first test we discuss is the *t-test for one sample*. In our statistical calculations, we assume that the observations in the sample are normally distributed. We use the following notation:

μ	Population mean
μ_0	Theoretical population mean
$\overline{X}$	Sample mean
σ^2	Population variance
s^2	Sample variance

(continued)

$s_{\bar{x}}^2$	Variance of the sample mean
$s_{\bar{x}}$	Standard error of the sample mean
t_α	Critical value for the t distribution
$\lvert t \rvert$	Test-statistic for the t distribution

We set up the hypothesis that the population mean is equal to the theoretical mean:

H_0: $\mu = \mu_0$
H_1: $\mu \neq \mu_0$
The test-statistic $t = \frac{\bar{x}-\mu_0}{S_{\bar{x}}}$

The test-statistic is the sample mean minus the theoretical mean, divided by the standard error of the sample mean. The critical cutoff value for t_α is found in the table for the t-distribution with degrees of freedom ($n - 1$). If the absolute value of the test-statistic $\lvert t \rvert > t_\alpha$, we reject H_0 and claim that the alternative hypothesis is supported. The logic of the t-test is to test the extent to which the mean value of the sample deviates from the theoretical value, taking into account the variance in the estimate of $s_{\bar{x}}$.

An important characteristic of the t-distribution with respect to hypothesis testing is that it is a symmetric bell-shaped curve with values that deviate from the mean in either direction (positive or negative). When stating a hypothesis based on a t-test, you can choose to not specify a direction, which is a *two-sided (or two-tailed) hypothesis*. For example, you state that employee turnover is different this year from the previous year without specifying whether it is higher or lower. If, however, you specify a direction, it is a *one-sided (or one-tailed) hypothesis*. For example, you state that employee turnover is higher this year than last year. You are not only testing whether it is different, you are testing that it is higher (not lower) than last year.

Example 9.1

Assume that the hotel industry in Scandinavia has, on average, 40% annual employee turnover. We want to test whether the hotels in our sample have the same turnover as the industry average. Since our sample could be higher or lower, this is a two-sided test. The null hypothesis and alternative hypothesis are:

H_0: $\mu_0 = 40\%$
H_1: $\mu_0 \neq 40\%$

Open the *Hotel* data. To estimate a one-sample t-test in SPSS, choose: *Analyze > Compare means > One-Sample T Test. Then move Emp_turnover to the Test Variable(s) box and put 40 into the Test Value box. Click OK* (see Fig. 9.2).

Fig. 9.2 One-sample *t*-test

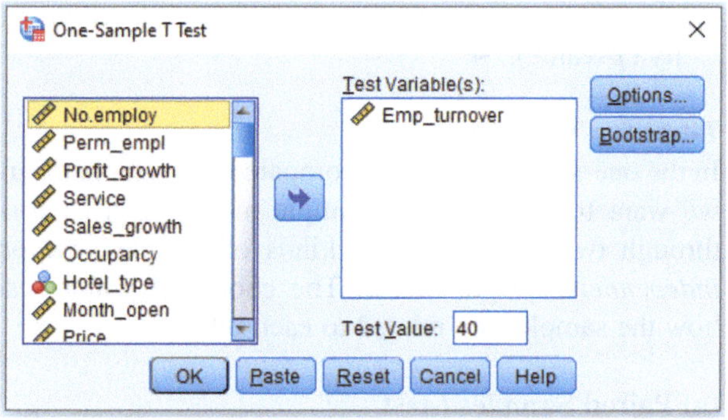

Table 9.2 One-sample *t*-test output

One-Sample Statistics

	N	Mean	Std. Deviation	Std. Error Mean
Emp_turnover	100	46.20	28.588	2.859

One-Sample Test

	Test Value = 40					
					95% Confidence Interval of the Difference	
	t	df	Sig. (2-tailed)	Mean Difference	Lower	Upper
Emp_turnover	2.169	99	.032	6.200	.53	11.87

In Table 9.2, the mean yearly employee turnover in the sample is 46.20, which exceeds the population mean (40%) by 6.2%. To check whether the difference is statistically significant, you need the degrees of freedom and the test-statistic for *t*. For a one sample *t*-test, the degrees of freedom are calculated by taking $n - 1$, which is 100–1 = 99. For your convenience, the degrees of freedom are shown in the lower table beside the *t*-value test-statistic. This is a two-sided test with alpha at 5%, so the critical value in the *t*-table is 1.984. The test-statistic is 2.169. Since $2.169 > 1.984$, we reject the null hypothesis and conclude that there is a significant difference between annual employee turnover in the sample compared to the population mean for hotels in Scandinavia. In other words, the sample has 6.2% higher employee turnover.

Since computers can calculate the *significance probability*, which is also called the *p*-value, we no longer need to look at the tables for hypothesis tests. The *p-value* for a two-sided test is reported as 0.032 in the lower table. Since $0.032 < 0.05$, the null hypothesis is rejected. Although we can see this directly from the output in most statistical software, it is still good practice to report the *t*-

value for the test-statistic and degrees of freedom, plus the significance probability (p-value). ◀

T-test for Two Samples

In the one-sample t-test, we compared the sample mean with a specific value. Often, we want to compare two sample means with each other. In this section, we go through two t-tests for two kinds of samples: the *paired samples t-test* and the *independent samples t-test.* The choice of which test is appropriate depends on how the samples are related to each other.

(a) Paired Samples t-test

In the *paired samples t-test*, the observations in one sample are directly related to the observations in the other sample, for example, the amount of sick leave from one year to the next for the same group of people at the same company. The sample is the same, measured at two time periods. The paired samples t-test is also called the *repeated measures t-test* or the *dependent samples t-test.*

The paired samples t-test is for testing mean differences in one population measured for the same thing at two time points. Imagine testing whether the labor costs for a company were equal for September one year μ_1 compared to September the next year μ_2. The hypotheses are:

H_0: $\mu_1 = \mu_2$
H_1: $\mu_1 \neq \mu_2$

First, note that this is a two-sided test since the alternative hypothesis is not specifying whether the mean is higher or lower for μ_2. It could be higher (one side) or lower (the other side). We simply test whether there is a difference in labor costs between the two time periods.

μ_1 and μ_2 are the mean values for the population measured in September year one and September year two. μ_1 and μ_2, and their associated variances σ_1^2 and variances σ_2^2, are assumed to be unknown. Therefore, we use the values from the samples taken at the two time periods. H_0 is tested by calculating the test-statistic using the following equation (assuming equal variance between samples) for a two-sided test. The test-statistic degrees of freedom are calculated by taking the sample size for the respective groups minus 1, $(n_1 - 1)$ and $(n_2 - 1)$. The variance is calculated by:

$$s^2_{(\bar{x}_1 - \bar{x}_2)} = \frac{\sum_i (x_{1i} - \bar{x}_1)^2 + \sum_i (x_{2i} - \bar{x}_2)^2}{(n_1 - 1) + (n_2 - 1)} \left[\frac{1}{n_1} + \frac{1}{n_2}\right]$$

Example 9.2

We test whether labor costs from September the first year are different from the labor costs in September the second year. Open the *Hotel* dataset. To calculate a

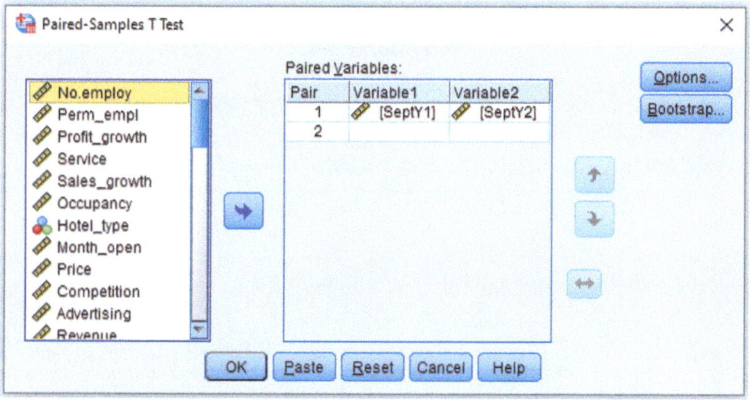

Fig. 9.3 Paired-samples *t*-test

Table 9.3 Paired-samples *t*-test: Means

Paired Samples Statistics

		Mean	N	Std. Deviation	Std. Error Mean
Pair 1	SeptY1	26787.00	100	11779.966	1177.997
	SeptY2	27814.00	100	12038.554	1203.855

Table 9.4 Paired-samples *t*-test: Correlation and *n*

Paired samples correlations

		N	Correlation	Sig.
Pair 1	SeptY1 & SeptY2	100	0.960	0.000

paired-samples *t*-test in SPSS, choose: *Analyze > Compare means > Paired-Samples T Test. Then, move SeptY1 and SeptY2 into the Paired Variables box. Click OK* (see Fig. 9.3).

As shown in Table 9.3, the mean labor costs for September of the first year were 26,787 kronor, while for September of the second year they were 27,814.

Table 9.4 shows that there are 100 observations for each year and that the correlation between years is 0.960.

Table 9.5, to the left side, shows that the mean difference is −1027. The question, however, is whether the mean difference between years is large enough to be statistically significant?

The test-statistic is $|-3.041|$. We are only concerned with the *absolute value* of the test-statistic (this time we included the absolute value notation $|t|$), so it does not matter whether the test-statistic is positive (+) or negative (−). With 99 degrees of freedom and a significance level of 5% for a two-sided test, the

Table 9.5 Paired-samples *t*-test: Significance test

Paired Samples Test

		Paired Differences					t	df	Sig. (2-tailed)
					95% Confidence Interval of the Difference				
		Mean	Std. Deviation	Std. Error Mean	Lower	Upper			
Pair 1	SeptY1 - SeptY2	-1027.000	3377.152	337.715	-1697.100	-356.900	-3.041	99	.003

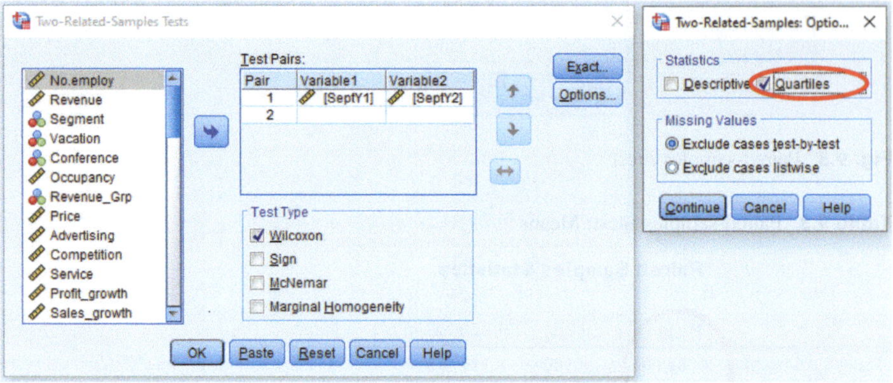

Fig. 9.4 Wilcoxon signed ranks dialogue box

critical value for *t* is 1.984. Since 3.041 > 1.984, we reject the null hypothesis and conclude that there is a significant difference for labor costs between years. Another way to reach this conclusion is to look at the significance probability. 0.000 (two-sided) is less than 0.05, so the null hypothesis is rejected. ◀

Nonparametric Wilcoxon Signed Ranks Test

If, for example, the continuous variables in the paired samples *t*-test were not normally distributed, then the *Wilcoxon signed ranks test* would be appropriate. It tests for median differences between paired samples or repeated measures by converting observations to ranks. We repeat the previous example, comparing labor costs between years. In the *Hotel* dataset, choose: *Analyze > Nonparametric Tests > Legacy Dialogs >2 Related Samples. Then, move SeptY1 and SeptY2 into the Test Pairs box. Click Options and tick Quartiles. Click Continue and OK* (see Fig. 9.4).

Table 9.6 shows the median for September of year 1 is 27,000 kronor, which is above the median for September of year 2, which is 25,205 kronor.

Table 9.7 shows that the difference between medians is statistically significant with the *p*-value of 0.001, which is less than 0.05. We draw the same conclusion as the paired samples *t*-test: that there is a significant difference between years. However, it is noteworthy that the difference is in the opposite direction. There is a

Table 9.6 Medians

Descriptive Statistics

	N	Percentiles 25th	50th (Median)	75th
SeptY1	100	20000.00	27000.00	35000.00
SeptY2	100	23000.00	25205.00	34500.00

Table 9.7 Wilcoxon signed ranks test

Test Statistics[a]

	SeptY2 - SeptY1
Z	-3.347[b]
Asymp. Sig. (2-tailed)	.001

a. Wilcoxon Signed Ranks Test

b. Based on negative ranks.

decrease in median labor costs between years. This demonstrates the importance of interpreting statistics as evidence. They do not prove one thing or another. They provide evidence for and against lines of reasoning, which are often represented as hypotheses. In this example, we would have to consider other variables to determine the direction of the trend in labor costs.

(b) Independent Samples *t*-test

The *independent samples t*-test determines whether there is a statistically significant difference between the means of two unrelated samples. The samples are assumed to be mutually exclusive, meaning that no case is present in both groups. A typical example is comparing gender (assuming two categories) for a continuous variable, like the amount of sick leave for men and women.

Example 9.3

Imagine that you want to test whether there is a difference between hotel types (boutique and chain) for how much annual revenue they earn. This is similar to the cross-tab example in Chap. 8, where we compared revenue groups across hotel types. In this example the hypotheses are:

H_0: $\mu_1 = \mu_2$
H_1: $\mu_1 \neq \mu_2$

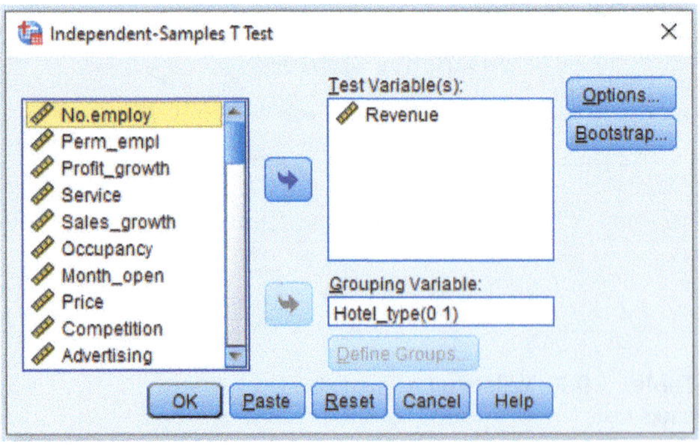

Fig. 9.5 Independent-samples *t*-test dialogue box

Table 9.8 Independent samples *t*-test output: Group statistics

Group Statistics

	Hotel_type	N	Mean	Std. Deviation	Std. Error Mean
Revenue	Boutique	40	15.45	16.934	2.678
	Chain	60	24.68	21.598	2.788

μ_1 is the mean annual revenue for boutique hotels, and μ_2 is the mean annual revenue for chain hotels. This is a two-sided test because we are not hypothesizing higher or lower revenue. Using the *Hotel* data, in SPSS choose: *Analyze > Compare means > Independent-Samples T Test. Then, move Sept18 into the Test Variable(s) box, and Hotel_type into the Grouping Variable box. You need to specify how the grouping variable is coded. As a rule of thumb, it is good practice to code dichotomous (two categories) variables as 0 and 1. Click on Define Groups, and put in 0 and 1. Click OK* (see Fig. 9.5).

In the Group Statistics table in Table 9.8, the mean annual revenue for boutique hotels is 15.45 million kronor and the mean annual revenue for chain hotels is 24.68 million kronor. There are 40 observations for boutique hotels and 60 observations for chain hotels. While there is no specific limit for how small the sample can be for a *t*-test, it is important that there are enough observations in each group to provide sufficient statistical power to detect effects. When the sample is too small, and thus power is too low, the likelihood of type II error increases. That is, not rejecting the null hypothesis when it is actually false. Put another way, the alternative hypothesis is not supported when it should be.

An assumption when comparing two group means in a *t*-test is that the variance between groups is equal. Since this is a common issue when comparing independent samples, the output in Table 9.9 shows *Levene's test of equality of variances*. The assumption is met when the associated *p*-value for the *F*-test is

Table 9.9 Independent samples *t*-test output: Levene and *t*-test

Independent Samples Test

		Levene's Test for Equality of Variances				
			Levene			
		F	Sig.	t	df	Sig. (2-tailed)
Revenue	Equal variances assumed	2.839	.095	-2.276	98	.025
	Equal variances not assumed			-2.388	95.339	.019

Table 9.10 Independent samples *t*-test effect size cutoff values

Effect size	Partial Eta Squared (η^2)
Small	0.01–0.059
Medium	0.06–0.137
Lange	0.138–1

Fig. 9.6 Independent samples *t*-test output: Levene and *t*-test

$$\frac{t^2}{t^2 + (N1 + N2 - 2)}$$

$$\frac{-2.276^2}{-2.276^2 + (40 + 60 - 2)}$$

$$= 0.050$$

Small effect
5% of variance

above 0.05. In Table 9.9, the Levene significance probability is 0.095, which is larger than 0.05, so the variances are assumed to be equal. When variances are assumed to be equal, interpret the statistics in the top row. When they are not assumed to be equal, interpret the statistics in the bottom row.

The *t*-value in the top row is −2-276, with 98 degrees of freedom. This is a two-sided test, so the critical cutoff value is 1.984. The absolute value of −2.276 > 1.984, so the null hypothesis is rejected, meaning that there is a statistically significant difference between boutique hotels and chain hotels for revenue. This can also be seen by looking at the significance probability (2-tailed), which is below 0.05.

In line with our discussion on theoretical and statistical significance, it is important to consider whether the significant difference between means found in the *t*-test is meaningful in practical terms. We know that sample size affects statistical power and that large samples increase the risk for rejecting the null hypothesis (supporting the alternative hypothesis). This is the same logic as when we discussed the absolute magnitude of correlation coefficients in Sect. 8.6 of

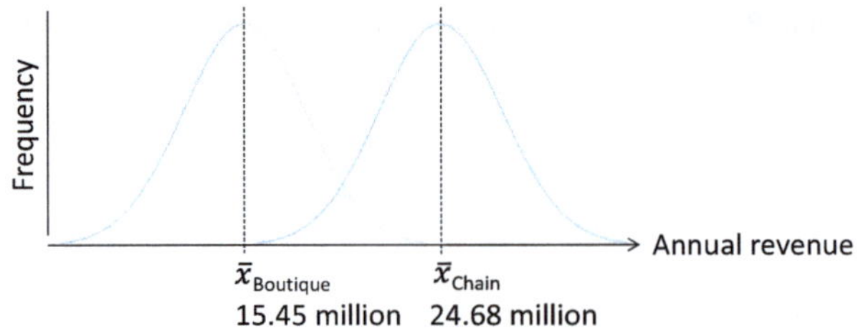

Fig. 9.7 Graph of independent samples *t*-test

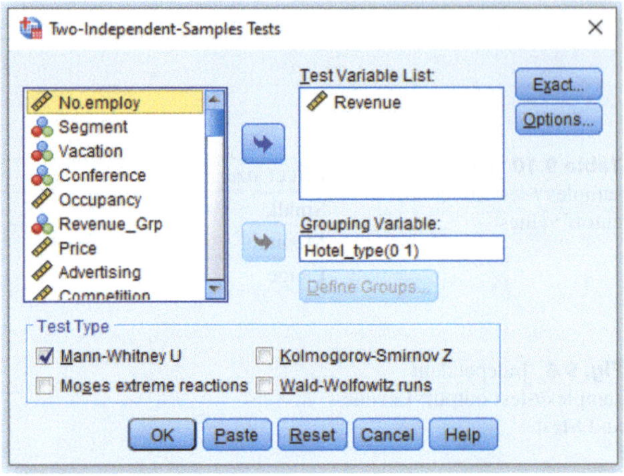

Fig. 9.8 Mann–Whitney U dialogue box

the previous chapter. There are several different measures of effect size when testing differences in group means. Table 9.10 shows cutoff values for interpretation. We show how to calculate Eta squared (η^2) as a measure of effect size in Fig. 9.6.

We conclude that the difference in annual revenue between boutique and chain hotels is statistically significant. However, the effect size is small. The graphical representation of the independent samples *t*-test for differences in wage costs is shown in Fig. 9.7. The normal curves and respective means represent the variance and degree of overlap between the distributions. Given the variance, the means are far enough apart for a statistically significant difference. ◀

Nonparametric Mann–Whitney U Test

The nonparametric alternative to the independent samples *t*-test is the *Mann–Whitney U test*. In the previous example, we tested for differences in the continuous variable, revenue, for two hotel types, boutique and chain. We checked the distribution of the *Revenue* variable, and found a skewness of 2.080, and a kurtosis of 4.924, which together indicate an abnormal distribution. The accepted cutoff indicating normality is an absolute value of 1. In this circumstance, it is advisable to test for group differences using a statistical method that is not based on having a normal distribution.

Using the *Hotel* data, choose *Analyze > Nonparametric Tests > Legacy Dialogs > Two Independent Samples. Put Revenue into the Test Variable List and Hotel type*

Table 9.11 Mann–Whitney mean rank

Ranks

	Hotel_type	N	Mean Rank	Sum of Ranks
Revenue	Boutique	40	40.19	1607.50
	Chain	60	57.38	3442.50
	Total	100		

Table 9.12 Mann–Whitney statistical significance

Test Statistics[a]

	Revenue
Mann-Whitney U	787.500
Wilcoxon W	1607.500
Z	-2.904
Asymp. Sig. (2-tailed)	.004

a. Grouping Variable: Hotel_type

into the Grouping Variable box. Define Groups as they are in the dataset, that is, 0 and 1. Click continue and OK (see Fig. 9.8).

Table 9.11 shows the mean ranks, indicating that boutique hotels at 40.19 million kronor are below chain hotels, at 57.38 million kronor.

Table 9.12, with a *p*-value of 0.004, which is below 0.05, indicates that the difference in mean ranks is statistically significant. We arrive at the same conclusion as with the independent samples *t*-test: that boutique hotels have a statistically significant lower revenue than chain hotels. T-tests are often called *robust against deviations from normality*. Given that the parametric *t*-test and nonparametric Mann–Whitney U-test arrive at the same conclusion, this is an example of the *t*-test being robust.

9.4 Analysis of Variance: One-way ANOVA

Analysis of variance (ANOVA) follows the same logic as the *t*-test, except that there are *three or more groups*. In the *Hotel* data, there is a variable called *Segment*. It has three categories: 1 for vacation hotels, 2 for conference hotels, and 3 for business hotels. It could be important for owners to understand which segment has higher occupancy rates. The *Occupancy* variable is the average of daily occupancy

Fig. 9.9 One-Way ANOVA

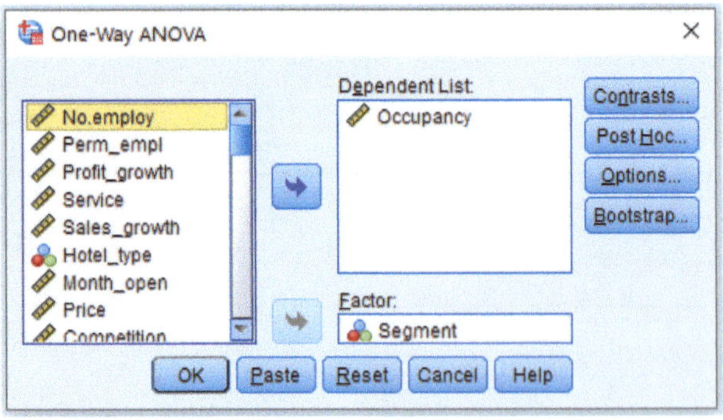

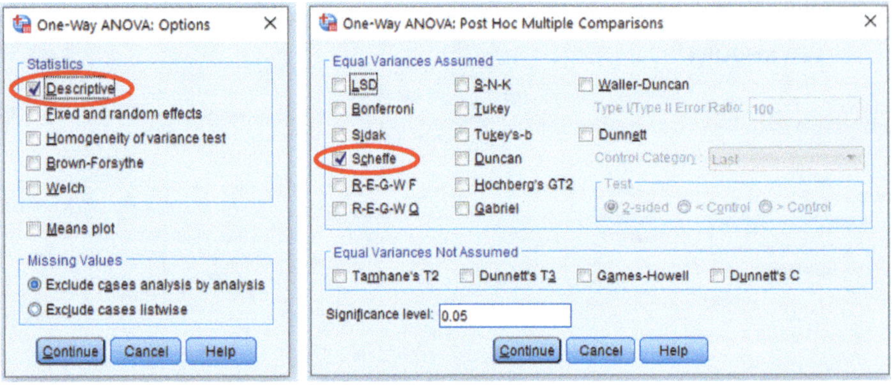

Fig. 9.10 ANOVA: Descriptives and post hoc tests

Table 9.13 ANOVA: Means for three groups

Occupancy	N	Mean	Std. deviation
Vacation	42	51.64	14.722
Conference	32	62.59	9.415
Business	26	64.77	12.359
Total	100	58.56	13.848

expressed in percent for the months open during one year. The t-test is limited to comparing only two sample means, whereas the F-test is appropriate for *comparing mean differences between three or more independent samples*. We show results here with a brief interpretation. In Chap. 10, we explain the F-test in greater detail and show how to test between multiple groups in regression. To estimate an ANOVA, open the *Hotel* dataset in SPSS. *Choose Analyze > Compare Means > One-Way ANOVA. Move Occupancy to the Dependent List and Segment to the Factor box* (see Fig. 9.9).

Choose Options, tick Descriptives, and then continue. Choose Post Hoc, then tick Scheffe, and continue (see Fig. 9.10). *Click OK.*

Table 9.13 shows the means for the three segments. Conference hotels and business hotels have similar occupancy rates at 62.59 and 64.77, respectively. Vacation hotels have a lower average occupancy at 51.64.

Table 9.14 ANOVA: *F*-test

ANOVA

Occupancy

	Sum of Squares	df	Mean Square	F	Sig.
Between Groups	3532.663	2	1766.332	11.088	.000
Within Groups	15451.977	97	159.299		
Total	18984.640	99			

Table 9.15 ANOVA: Post hoc tests

Multiple Comparisons

Dependent Variable: Occupancy
Scheffe

(I) Segment	(J) Segment	Mean Difference (I-J)	Std. Error	Sig.	95% Confidence Interval Lower Bound	95% Confidence Interval Upper Bound
Vacation	Conference	-10.951*	2.962	.002	-18.31	-3.59
	Business	-13.126*	3.150	.000	-20.96	-5.30
Conference	Vacation	10.951*	2.962	.002	3.59	18.31
	Business	-2.175	3.332	.808	-10.46	6.11
Business	Vacation	13.126*	3.150	.000	5.30	20.96
	Conference	2.175	3.332	.808	-6.11	10.46

*. The mean difference is significant at the 0.05 level.

The *F*-test in Table 9.14 indicates that there is a significant difference between groups. This is indicated by the significance probability, which is less than 0.05.

Table 9.15 shows the post hoc tests. The *F*-test indicated that there is a significant difference between groups. However, it does not indicate where the differences may be. There could be, for example, only one significant difference between two groups, or significant differences between all groups. The significance probabilities in the red ovals show that there is a significant difference for occupancy between Vacation and Conference hotels ($0.02 < 0.05$), and Vacation and Business hotels ($0.00 < 0.05$). There is no significant difference between Business and Conference hotels for occupancy ($0.808 > 0.05$). Note that we chose the Scheffe post hoc test. There are several types of post hoc tests with different strengths and weaknesses. A discussion of post hoc tests is beyond the scope of this text.

An alternative way of testing between groups for a continuous dependent variable is to use OLS regression. In the current example with three categories, you would form two dummy variables. The dummy variables would be input as independent variables, and *Occupancy* would be the dependent variable. We will demonstrate this in Chap. 10 when we explain regression.

9.5 Chi-square Test (Chi²)

Notation for the discussion of the chi-square test:

Chi^2	Chi-square
Corr	Pearson correlation coefficient
SR	Spearman's rank correlation coefficient
O_i	Observed frequency
E_i	Expected frequency
v	Degrees of freedom
χ^2	Test-statistic for the chi-square distribution
χ^2_α	Critical cutoff value for the chi-square distribution

Relationships between variables and dependency of variables on each other is a central theme for many disciplines within management. In this section, we discuss *cross-tabulation* and the *chi-square test for independence*, which is appropriate for testing relationships between two variables with two or more nominal level categories. An assumption of the chi-square test is that each cell (group of data) contains at least five observations.

Frequency Differences

In Chap. 8, we showed how to generate a cross-tabulation table for hotel type (boutique or chain) and revenue group (low or high). This is a classic 2 by 2 design. Table 9.16 shows the output of the cross-tabulation with percentages added for rows. Apparently, chain hotels are quite equally distributed across revenue groups (53.3% low and 46.7% high). Boutique hotels are mostly in the low revenue group (85%). As a rule of thumb, we put the variable for which we want to compare something in the column position. The classic example is testing whether there is a difference between male and female (gender variable in row), for being smokers or not

Table 9.16 Cross-tabulation for hotel type and revenue group

Hotel_type * Revenue_Grp Crosstabulation

			Revenue_Grp		
			Low	High	Total
Hotel_type	Boutique	Count	34	6	40
		% within Hotel_type	85.0%	15.0%	100.0%
	Chain	Count	32	28	60
		% within Hotel_type	53.3%	46.7%	100.0%
Total		Count	66	34	100
		% within Hotel_type	66.0%	34.0%	100.0%

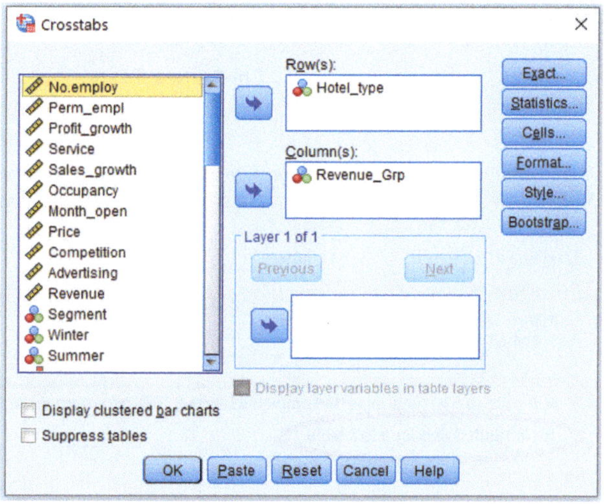

Fig. 9.11 Cross-tabs for hotel type and revenue group

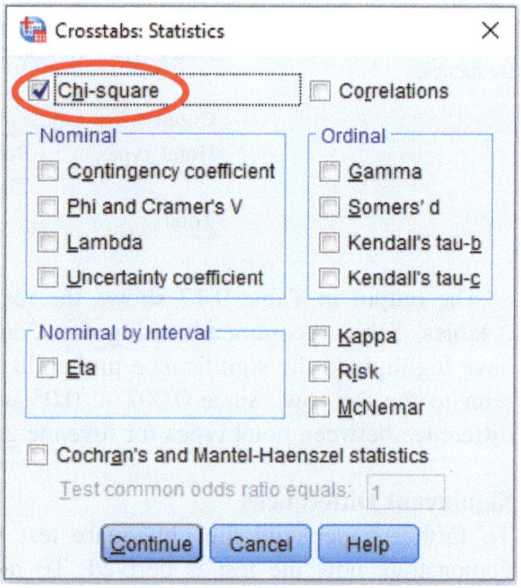

Fig. 9.12 Cross-tabs: Chi-square test

(smoking variable in column). Another way to think about it is: smoking does not cause gender, whereas your gender can determine the likelihood of smoking.

With the chi-square statistic, we can test whether observed differences are statistically significant. For the hotel example, the hypothesis could be expressed as: on average, boutique hotels are far more likely to be classified in the low revenue group than chain hotels. To generate a chi-square test, in SPSS, open the *Hotel* dataset. Choose: *Analyze > Descriptive Statistics > Crosstabs > move Hotel_type to Row(s) and Revenue_Grp to Column(s)* (see Fig. 9.11).

Then, choose Statistics and tick the Chi-Square box (see Fig. 9.12). *Then, Continue and OK.*

Table 9.17 Cross-tabs: Chi-square test

Chi-Square Tests

	Value	df	Asymptotic Significance (2-sided)	Exact Sig. (2-sided)	Exact Sig. (1-sided)
Pearson Chi-Square	10.725[a]	1	.001		
Continuity Correction[b]	9.360	1	.002		
Likelihood Ratio	11.480	1	.001		
Fisher's Exact Test				.001	.001
Linear-by-Linear Association	10.618	1	.001		
N of Valid Cases	100				

a. 0 cells (0.0%) have expected count less than 5. The minimum expected count is 13.60.

b. Computed only for a 2x2 table

Table 9.18 Cross-tab cell frequencies

Hotel_type * Revenue_Grp Crosstabulation

Count		Revenue_Grp Low	High	Total
Hotel_type	Boutique	34	6	40
	Chain	32	28	60
Total		66	34	100

The output in Table 9.17 shows the results of the Chi-square test. For 2 by 2 tables, SPSS recommends using the Continuity Correction, which is why we have highlighted the significance probability in the second row. For larger tables, refer to the top row. Since $0.002 < 0.05$, we conclude that there is a significant difference between hotel types for revenue group.

Significant Differences

To further understand the chi-square test for independence, in this section we demonstrate how the test is derived. To facilitate the comparison, we show the output again, with just cell frequencies in Table 9.18.

1. Hypothesis Formulation

H_0: There is no difference between groups (they are independent)

H_1: There is a difference between groups (they are dependent)

Usually, a 5% significance level is used with the Chi-square test.

2. Calculate the Expected Frequency
Calculate the expected value E_i for every cell in the table, assuming that there is no dependency between the variables. For the first cell in the table (boutique and low revenue), the formula is:

$$E_i = \frac{row\ sum * column\ sum}{total\ sum} = \frac{40 * 66}{100} = 26.4$$

The expected frequency E_i is 26.4, while the observed frequency O_i (the value in the cell) is 34. Continue and do this for every cell.

3. Calculate the Squared Difference (for Cell 1)
Calculate the squared difference between the observed frequency O_i and the expected frequency E_i for each cell. For the first cell it is:

$$Cell\ 1 = \frac{(O_i - E_i)^2}{E_i} = \frac{(34 - 26.4)^2}{26.4} = 2.19$$

The relative squared difference between the observed frequency O_i and the expected frequency E_i for cell 1 is 2.19.

4. Calculate Chi-square Test-Statistic
The sum of the relative *squared differences* is the chi-square test-statistic. It is calculated according to the following formula:

$$\chi^2 = \frac{(O_i - E_i)^2}{O_i} = \frac{(34 - 26.4)^2}{26.4} + \frac{(32 - 39.6)^2}{39.6} + \frac{(6 - 13.6)^2}{13.6} + \frac{(28 - 20.4)^2}{20.4}$$
$$= 10.72$$

Chi-square is 10.72. The greater the differences between the observed frequencies O_i and the expected frequencies E_i, the larger chi-square will be. The chi-square test-statistic can be compared with the critical value in the chi-square table for the chi-square distribution. However, we need to calculate the degrees of freedom according to the following formula:

$$v = (sum\ of\ rows - 1) * (sum\ of\ columns - 1) = (2 - 1) * (2 - 1) = 1$$

With 1 degree of freedom and a significance level of 5%, we refer to the chi-square table to find the critical cutoff value $\chi_\alpha^2 = 3.84$. Since the test-statistic $\chi^2 = 10.72$, and $10.72 > 3.84$, we reject the null hypothesis and conclude that there is a dependency between the variables. That is, boutique hotels are significantly more likely to be classified in the low revenue group. This is in line with our conclusions from Table 9.13.

9.6 Testing Correlation Coefficients

In Chap. 8, we described two correlation coefficients: the Pearson correlation (*Corr*) and Spearman's rank correlation (*SR*). We now see how to test whether the correlation coefficient between two variables *X* and *Y* is statistically significantly different from zero. In this section, we will use the following notation:

ρ	Population value
X	Variable X
Y	Variable Y
Corr	Pearson correlation coefficient
SR	Spearman's rank correlation coefficient

Pearson Correlation

The most common type of correlation coefficient is the *Pearson correlation*. We test the null hypothesis H_0 that there is no correlation (*Corr*) between variable *X* and variable *Y*, against the alternative hypothesis H_1 that there is a correlation between *X* and *Y*. We write this as a hypothesis:

H_0: $\rho = 0$
H_1: $\rho \neq 0$

The test-statistic, which follows the t-distribution, can be calculated by the following formula:

$$t = \frac{Corr(X, Y)}{\sqrt{\frac{1-Corr(X,Y)^2}{n-2}}}$$

The *degrees of freedom* (v) are the number of observations minus two ($n - 2$). The null hypothesis is rejected if the test-statistic $|t|$ is greater than the critical value t_α, found in the *t*-tables.

In Chap. 8, we estimated the correlation between the number of employees at a hotel and the hotel's revenue. In SPSS using the *Hotel* data, choose: *Analyze > Correlate > Bivariate > move No.employ and Revenue to the Variables box and click OK*. Now, we are considering whether the correlation coefficient is statistically significant. The significance level highlighted in Table 9.19 is less than the standard 0.05 cutoff ($0.00 < 0.05$), so we reject the null hypothesis and consider the correlation coefficient (0.924) to be statistically significant. SPSS also provides asterisk* notation flagging significant correlations.

SPSS and most statistical software programs *run a two-sided (two-tailed) test by default*. This means that no direction is specified in the hypothesis (positive or negative). To get the one-sided significance (*p*-value), simply divide the significance in half. In this case, it is so small (0.000) that there is no need.

Table 9.19 Pearson correlation statistical significance

Correlations

		No.employ	Revenue
No.employ	Pearson Correlation	1	.924**
	Sig. (2-tailed)		.000
	N	100	100
Revenue	Pearson Correlation	.924**	1
	Sig. (2-tailed)	.000	
	N	100	100

**. Correlation is significant at the 0.01 level (2-tailed).

We can also calculate the test-statistic for the t-distribution. Starting with the number of observations (100), we calculate the degrees of freedom v by taking $100 - 2 = 98$. We insert these values into the following formula to find the test-statistic:

$$t = \frac{Corr(X,Y)}{\sqrt{\frac{1-Corr(X,Y)^2}{v}}} = \frac{0.924}{\sqrt{\frac{1-0.924^2}{98}}} = 23.92$$

According to the table for the t-distribution, the critical cutoff value for $n = 98$ is 1.984 at α of 0.05 (i.e., 5% significance level two-sided test). Since $|t| > t\alpha$ $(23.92 > 1.984)$, we reject the null hypothesis. This supports our conclusion that there is a significant correlation between the number of employees and revenue at the hotels in our sample.

Spearman's Rank Correlation

In Chap. 8, we showed how to estimate the nonparametric correlation coefficient, *Spearman's rank correlation*. We continue the example here and show how to test the statistical significance of the correlation. To estimate the Spearman's rank correlation with the *Hotel* data for the variables Profit_growth and Sales_growth. Choose: *Analyze > Correlations > Bivariate > and tick the box for Spearman*. The output in Table 9.20 shows the correlation matrix with the significance probability highlighted.

Spearman's rank correlation is considered as a special type of Pearson correlation, so the calculation of the test-statistic is essentially the same. This gives the following equation:

$$t = \frac{SR(X,Y)}{\sqrt{\frac{1-SR(X,Y)^2}{v}}} = \frac{0.435}{\sqrt{\frac{1-0.435^2}{98}}} = 4.78$$

According to the t-table, the critical cutoff value for a two-sided test with n = 98 is 1.984 at α of 0.05. Since $4.78 > 1.984$, we reject the null hypothesis. This supports

Table 9.20 Spearman correlation statistical significance

Correlations

			Profit_growth	Sales_growth
Spearman's rho	Profit_growth	Correlation Coefficient	1.000	.435**
		Sig. (2-tailed)	.	.000
		N	100	100
	Sales_growth	Correlation Coefficient	.435**	1.000
		Sig. (2-tailed)	.000	.
		N	100	100

**. Correlation is significant at the 0.01 level (2-tailed).

Table 9.21 Nonparametric alternatives

Parametric	Non-parametric
Pearson correlation	Spearman rank correlation
Paired samples *t*-test	Wilcoxon test
Independent samples *t*-test	Mann-Whitney U-test
ANOVA related samples (repeated measures)	Friedman test
ANOVA independent samples (between groups)	Kruskal-Wallis test

our conclusion that there is a significant correlation between sales growth and profit growth at the hotels in our sample.

9.7 Summary

In statistical terms, hypothesis testing is about rejecting or retaining the null hypothesis. Whether the null hypothesis is supported or rejected is determined by the selected significance level and the relevant test-statistic. The test-statistic is calculated differently according to the probability distribution on which the test is based. For example, we discussed the *t*-test, the *F*-test, and the Chi-square test, which are all based on their respective probability distributions. What is crucial in any significance test is whether the test-statistic deviates more from zero than we would expect if the null hypothesis were supported. This is determined by comparing the test-statistic with critical values for the relevant probability distribution. Critical values are listed in tables in many statistics books and online.

Table 9.21 shows a list of parametric tests and their nonparametric alternatives.

Regression Analysis 10

Contents

10.1 Introduction

Imagine that someone has a lot of data about you: where you work, go to school, what you like, what you do not like, who your friends are, and so on. Then, based on this data, with great accuracy they can predict your behaviors and choices. What car you will purchase, where you will go on holiday, what you will buy your mother for her birthday. They can do this because most human behaviors are rational in so far as they follow consistent patterns; we are creatures of habit. And, they can predict your behavior because they know how to use regression analysis.

In this chapter, we explain regression analysis. It is the most classic and widely used statistical method for analyzing the dependency relationship between two or

J. E. Sallis et al., *Research Methods and Data Analysis for Business Decisions*, Classroom Companion: Business, https://doi.org/10.1007/978-3-030-84421-9_10

more variables. *Regression analysis* estimates the effect of one or more independent variables (*X*) on a single dependent variable (*Y*). We use the following notation:

X	Variable X (independent variable)
Y	Variable Y (dependent variable)
β_1	Regression parameter
β_0	The constant (regression intercept)
$\widehat{Y}$	The predicted value of Y
$\widehat{\beta}$	The estimate of beta
ε_i	The error term (also called the disturbance term)
e_i	The estimate of the error term (alternative notation: $\widehat{\varepsilon}_i$)
R^2	Explained variance of a regression equation (coefficient of determination)
$\overline{R}^2$	R-square adjusted. Also expressed as R^2_{adj}
TSS	Total sum of squares (total variance)
RSS	Residual sum of squares (residual variance)
ESS	Explained sum of squares (explained variance)
F	Test-statistic for the F distribution
H_0	Null hypothesis
H_i	Alternative hypothesis, where i is any whole number (1, 2, 3, . . .i)
n	Sample size
OLS	Ordinary lease squares estimation
$\|t\|$	Absolute value of the test-statistic for the t-distribution
Corr (X, Y)	Correlation between variable X and variable Y
k	k independent variables (X_1, X_2, . . . X_k)
i	The ith observation

Regression analysis is one of many statistical methods used to evaluate the relationship between one or more so-called *independent variables* X_1, X_2, . . . X_k and a *dependent variable Y*. Specifically, regression analysis assesses how changes in the independent variables explain changes in the dependent variable. For example, regression can be used to study:

1. Patient satisfaction (Y), and the independent variables physician communication (X_1) and service level during hospital stay (X_2).
2. Demand for a product (Y), and the independent variables price (X_1), advertising (X_2), and market share (X_3).
3. Change in blood pressure (Y), and hypertension medication (X_1).
4. Customer loyalty (Y), and price (X_1), customer satisfaction (X_2), and product quality (X_3).
5. Crop yield (Y), and rainfall (X_1), air temperature (X_2), and soil nitrogen content (X_3).

While these examples imply a cause–effect relationship from the independent (X) variables to the dependent (Y) variable, regression itself does not prove causality. It simply tests whether the modeled correlations are significantly different from zero.

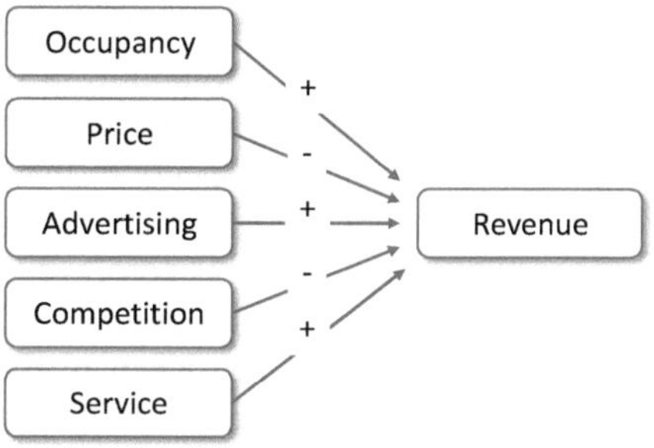

Fig. 10.1 Research model for factors affecting hotel revenue

For example, we believe that occupancy rates, price, advertising, competition, and service level determine the amount of revenue at a hotel (see Fig. 10.1). We cannot prove this, but with regression, we can test for significant relationships between the five independent variables and the dependent variable, revenue. If we find significant relationships, we can predict revenue for different values of each independent variable. That is, how much would a change in occupancy, price, advertising, competition, or service change revenue. Causality is based on robust argumentation and theory and supported (or refuted) by evidence.

We expect that occupancy, advertising, and service will have a positive effect on revenue, while price and competition are expected to have a negative effect. This means that if the occupancy rate goes up, advertising is increased, or service is improved, then revenue will increase. Whereas if the price goes up or the number of competitors increases, revenue will decrease. The choice of independent variables should be logical, and if possible, theoretically justified. We mean that it should be possible to argue for each variable based on common sense, and when appropriate, established theory.

Most often, the relationship between the dependent and independent variables is assumed to be linear (a straight line). In our example, this means that revenue is assumed to be a linear function of the independent variables. It is important to understand that the five independent variables do not account for everything that will cause revenue to rise or fall. To account for all the other possible things that could affect revenue, we include an error term in the regression equation. It is often symbolized by the Greek letter epsilon (ε). This is the theoretical regression model for the model in Fig. 10.1:

$$\begin{aligned} Revenue = {} & \beta_0 + \beta_1 Occupancy + \beta_2 Price + \beta_3 Advertising + \beta_4 Competition \\ & + \beta_5 Service + \varepsilon \end{aligned}$$

Note that we use plus signs (+) when expressing the general regression equation, even though we expect price and competition to have a negative relationship with revenue. When the beta coefficients are estimated, if they are negative, they will have the proper relationship with the dependent variable (a negative and a positive equal a negative).

β_0 (beta null) is called the *constant* or sometimes the *Y-intercept*. It indicates where the regression line intercepts the Y-axis in a two-dimensional plane. The *constant* β_0 is what the value of the dependent variable Y (revenue) would be if all the independent variables (occupancy, price, advertising, competition, and service) were equal to zero. Often, the constant is pragmatically meaningless, only making sense from a mathematical perspective. Imagine using height to predict a person's body weight. You collect data from 50 respondents measuring their height and weight. In a regression, when height is zero, weight will almost certainly be a nonzero number on the Y-axis. In reality, if height could be zero (which is impossible), weight could not be anything but zero. Keep in mind that you must include the constant in a regression equation, even when it does not have practical meaning.

The betas (β_i, *where* $i = 1, 2, 3, 4, 5$) in front of each independent variable are called the beta coefficients, and ε is the error term. The right side of the equation (revenue is on the left side) can be divided into two parts. The error term represents the unexplained part, and the independent variables represent the explained part. Note, that even though we call it the error term, it does not necessarily refer to errors. Sometimes, it is called the disturbance term or the residual term. It must be included in the equation to represent the variance that is left over (i.e., not explained by the independent variables). The ambition with regression is to have the unexplained part ε as small as possible. In other words, we want a regression equation with as much *explanatory power* as possible. R^2 is a measure of the explained variance in regression.

The regression coefficients β_i ($i = 1, 2, 3, 4, 5$) indicate the "isolated" effect that each independent variable has on the dependent variable, revenue. For example, all else equal, β_3 will tell how much effect a one-unit change in advertising will have on revenue. *All else equal* means that we assume the other independent variables (occupancy, price, competition, and service) are kept constant. If, for example, the beta coefficient β_3 turns out to be 2.09, and advertising and revenue are measured in millions of kronor (the unit of measurement), then increasing advertising efforts by one million kronor will result in an expected revenue increase of 2,090,000, provided that the other variables are not changed. This is the *predicted change* in revenue.

Regression is common across many subjects and disciplines: finance, medicine, psychology, agriculture, and even sports. Movie tip: Based on a true story, watch the movie *Moneyball* with Brad Pitt and Jonah Hill. Billy Beane (Pitt) hires Paul DePodesta (Hill) to predict which over-valued players to trade away and which under-valued players to acquire. Based on solid data analysis, including regression, they built an inexpensive baseball team that won a record 20 straight games in the American baseball league's 2002 season. Arguably, they changed how baseball is managed through data analysis.

10.2 Simple Regression Analysis

Imagine a stochastic (random) process, which includes a dependent variable Y and several independent variables $X_1, X_2, X_3, \ldots, X_k$. We could write $Y = F(X_1, X_2, X_3, \ldots X_k) + \varepsilon$, which means that Y is a function of the independent variables

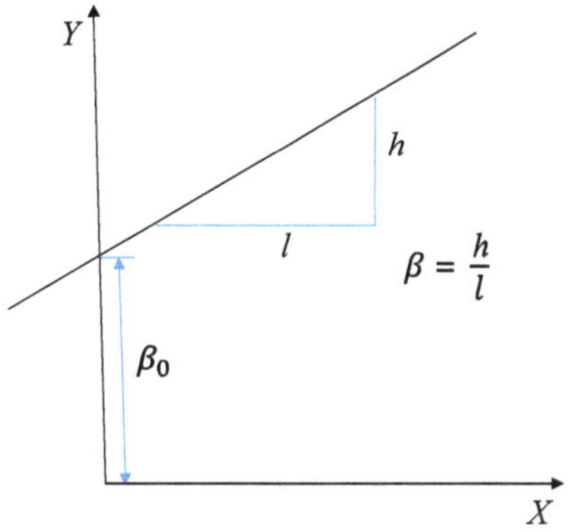

Fig. 10.2 Simple regression line

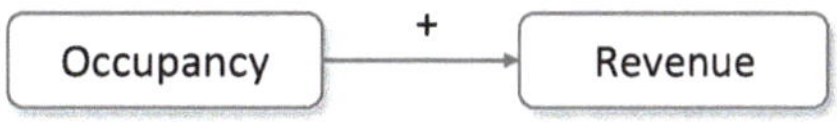

Fig. 10.3 Simple regression model

(X_i) plus the random term. Given the extreme unlikelihood of knowing all the independent variables in a function, and the unlikelihood of having a perfect set of population data, the true model is most often unknown. Instead, we try to identify the most important independent variables and measure them (plus the dependent variable) in a sample. In its simplest form, *simple regression analysis* estimates the relationship of one independent variable on one dependent variable. The theoretical model may be expressed as follows:

$$Y = \beta_0 + \beta_1 X + \varepsilon$$

Where Y is the dependent variable and X is the independent variable. β_0 is the constant (also the Y-axis intercept) and β_1 is the slope of the regression line. ε represents the error term. The explained part $\beta_0 + \beta_1 X_1$, can be represented graphically as a straight line in a coordinate system where the constant β_0 is the Y-intercept for the line and β_1 is the slope of the line (see Fig. 10.2). The beta parameters β_0 and β_1 are what we want to estimate with the regression equation. Given that we are not including X_2, X_3, X_4,. . .X_k, in the equation, their missing contribution to the equation is captured by the error term ε. Other variation in the error term can come from measurement error, an incorrect functional form (e.g., should the regression line actually be curvilinear), or pure coincidence.

10.3 Estimating Regression Parameters

We use the *Hotel* data to show an example of simple regression (see Fig. 10.3).

The hypothesis could be written: H_1: Occupancy has a positive effect on revenue.

In other words, we believe that the more rooms occupied in the hotel, the higher the hotel's revenue, measured in kronor. In general, when writing hypotheses in

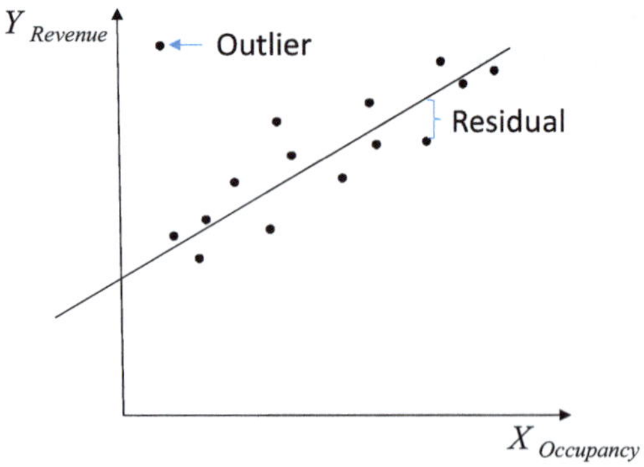

Fig. 10.4 Simple regression line with residuals and outlier

research reports, the alternative hypothesis is presented and the null hypothesis is implicit.

With the *Hotel* data, imagine that managers have been asked to provide occupancy data and revenue data. Occupancy and revenue are unlikely to be perfectly related. That is, a specific occupancy does not lead to exactly the same revenue in every hotel. Figure 10.4 shows how the data (the dots) may be represented in a two-dimensional plane. The data in this example are randomly spread in a somewhat linear fashion with a positive slope. The relationship between the two variables can be represented by the regression line that intersects them. The goal of the regression is to estimate the most efficient line intersecting the data. We are demonstrating ordinary least squares (OLS) regression, which is one of the most common estimation methods. The residuals are the distances from the data points to the regression line. They represent the error (unexplained variance) in the equation. If the data is more spread, the resdiduals get bigger, and the regression has more unexplained variance. We have also included one piece of data called an outlier. This would be a hotel that, despite very low occupancy, has very high revenue. Outliers often have an inordinately large influence on the parameter estimates, so they must be evaluated and possibly removed from the data. Perhaps the outlier is a luxury hotel catering to a few very wealthy clients. You would have to ask yourself whether it is appropriate to it keep in the dataset. If, for example, you are trying to draw general conclusions about typical hotels, it may be best to remove the luxury hotels from the sample.

To run the regression analysis in SPSS with the *Hotel* data, choose: *Analyze > regression > Linear. Put Revenue in the Dependent box and Occupancy in the Independent(s) box, Click OK* (see Fig. 10.5).

The Model Summary table (see Table 10.1) shows R^2 and R^2_{adj}. They are sometimes referred to as the *coefficient of determination.* They represent two measures of the explained variance in the regression equation. R^2_{adj} is based on R^2, with the additional element of taking into account the number of independent variables. *Parsimony* is a virtue in regression models. In general, you should aim to find the simplest possible regression model to explain the most amount of variance in the dependent variable. From this perspective, adding more independent variables to explain slightly more variance is counterproductive. R^2_{adj} includes a

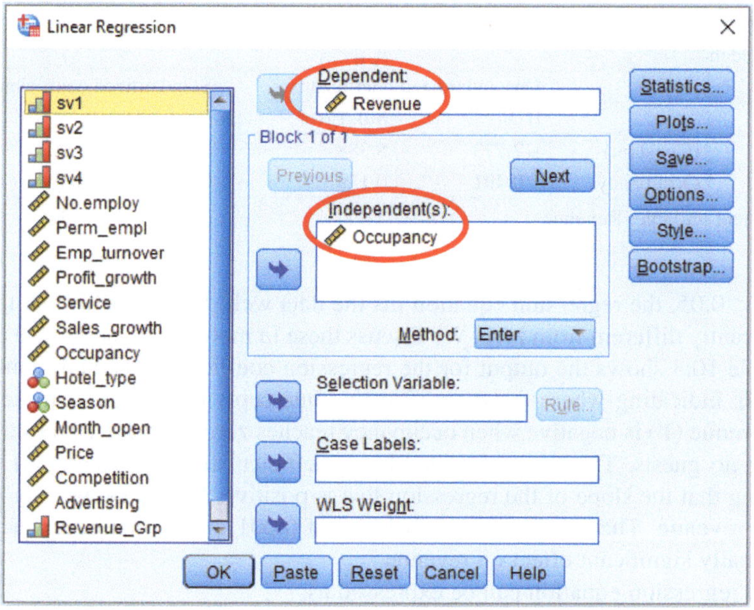

Fig. 10.5 Regression dialogue box

Table 10.1 Model summary: R^2 and R^2_{adj}

Model summary				
Model	R	R square	Adjusted R square	Std. error of the estimate
1	0.335[a]	0.112	0.103	19.212

[a]Predictors: (Constant), Occupancy

Table 10.2 ANOVA: *df*, *F* & Sig

ANOVA[a]						
Model		Sum of squares	df	Mean square	F	Sig.
1	Regression	4579.882	1	4579.882	12.408	0.001[b]
	Residual	36,173.108	98	359.113		
	Total	40,752.990	99			

[a]Dependent variable: Revenue
[b]Predictors: (Constant), Occupancy

penalty for each additional independent variable. If the contribution to the explained variance is minimal, R^2_{adj} will drop with the addition of the variable. The table shows a fairly low R^2 of 11.2%. This means that 88.8% of the variance is unexplained.

Table 10.2 shows the ANOVA table output. The *degrees of freedom* (*df*) are used for referring to the *t*-tables and *F*-Tables. *F* shows the test-statistic for the *F*-tables. Sig. is the significance probablility, which is also referred to as the p-value. Since

Table 10.3 Regression coefficients

Coefficients[a]						
		Unstandardized coefficients		Standardized coefficients		
Model		B	Std. Error	Beta	t	Sig.
1	(Constant)	−7.773	8.388		−0.927	0.356
	Occupancy	0.491	0.139	0.335	3.522	0.001

[a]Dependent variable: Revenue

0.001 < 0.05, the regression equation fits the data well. It also means that the R^2 is significantly different from zero. We discuss these in more detail later in the chapter.

Table 10.3 shows the output for the regression coefficients. The constant (β_0) is −7.726, indicating where the regression line intercepts the *Y*-axis. It makes sense that revenue (*Y*) is negative when occupancy reaches zero. Hotels have costs despite having no guests. The unstandardized beta coefficient for occupancy (β_1) is 0.49, meaning that the slope of the regression line is positive. Higher occupancy leads to higher revenue. The significance probability of 0.001 means that occupancy has a statistically significant effect on revenue.

The regression equation can be expressed as:

$$Y_i = -7.773 + 0.491X_1 + e_i$$

Which is the estimated model for observation *i*, where Y = Revenue and X = Occupancy. A one-unit change in occupancy (one room rented) will increase revenue with 0.49 kronor (measured in thousands).

We can create a plot of the regression line by choosing: *Graphs > Chart Builder > then choose (1) Scatter/Dot, then (2) drag the simple scatterplot up to the open box, then (3) drag Revenue to the Y box and Occupancy to the X box. Click OK* (see Fig. 10.6).

Double click on the graph in the output and then click on the *fit line at total* icon (see Fig. 10.7). This adds the regression line to the graph (see Fig. 10.8). Finally, to get the proper perspective, *click on the X and make the minimum zero, and click on Y and make the minimum − 10.*

In Fig. 10.8, you can see the regression line for predicting values of revenue. The *Y*-intercept (constant β_0) is −7.77 and the slope (β_1) of the line is 0.49. Compared to the example in Fig. 10.4, the data is more randomly spread around the graph. We have marked one of the residuals to show that in some cases they are very large. Accounting for the size of the residuals, it is not surprising that the explained variance (R^2) is only 11.2%.

The most common estimation method for regression is *ordinary least squares* (OLS) estimation. The idea is to find the estimates where the sum of the squared residuals is as small as possible. The formulas for the beta estimates are:

$\widehat{\beta} = \frac{\sum xy}{\sum x^2}$ and $\widehat{\beta}_0 = \overline{Y} - \widehat{\beta}\overline{X}$

Defined this way, *x* and *y* are often called the *deviation scores*.

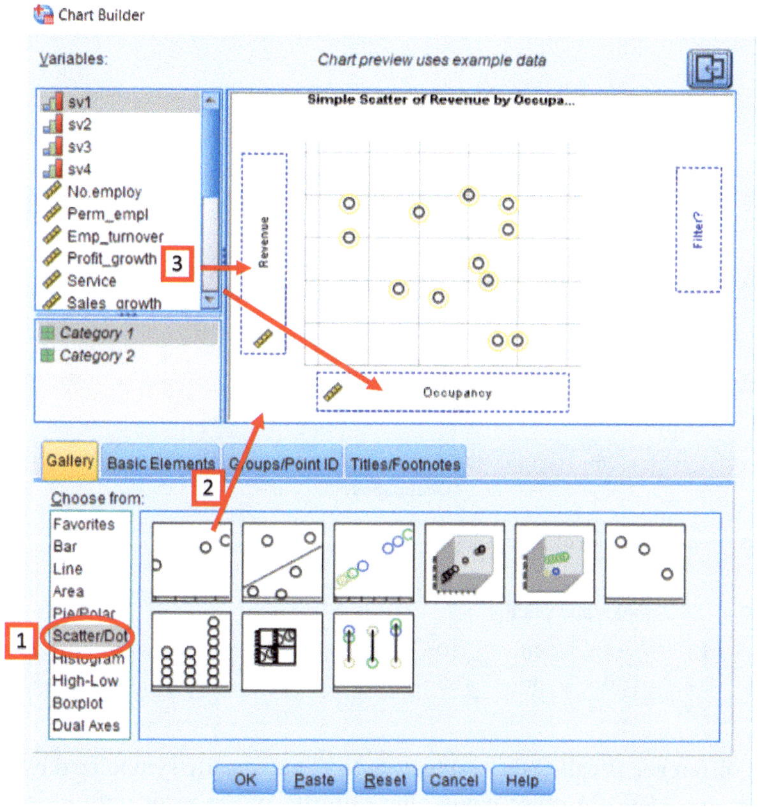

Fig. 10.6 Regression scatterplot

Fig. 10.7 Fit line at total icon

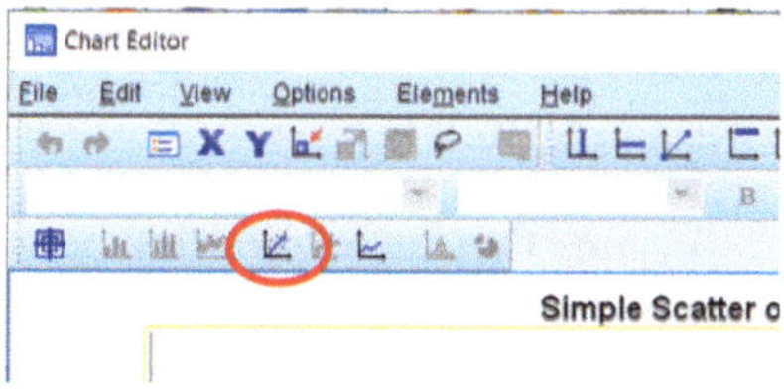

$$x = X - \overline{X} \text{ and } y = Y - \overline{Y}$$

Where $\overline{X}$ and $\overline{Y}$ are the respective mean values. Visually, this means determining the slope of the regression line so that the sum of the squared residuals (estimated error) is minimal. The error term (ε) can be calculated/estimated by the formula:

$$e_i = Y_i - \left(\widehat{\beta}_0 + \widehat{\beta} X_i\right)$$

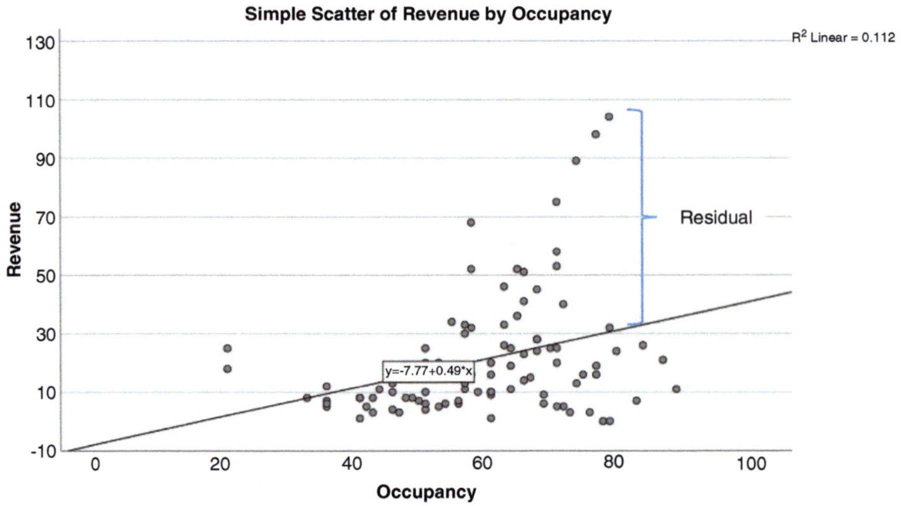

Fig. 10.8 Regression plot of occupancy on revenue

Table 10.4 Sold apples and price

Price	12	14	10	10	15	15	20	15	20	20
Sales	34	30	36	35	22	21	25	30	20	16

This difference is called the residual. $\widehat{\beta}_0 + \widehat{\beta}_i$ are usually symbolized by $\widehat{Y}_i$, such that $\widehat{Y}_i = \widehat{\beta}_0 + \widehat{\beta}X_i$. In other words, the estimate of the error term e_i is equal to Y minus the estimate for $\widehat{Y}$, which is $e_i = Y_i - \widehat{Y}_i$. Note that we have shown e_i to specifically mean that it is the *predicted* or *estimated* error value. An alternative notation would be $\widehat{\varepsilon}$.

Example 10.1

Assume that the relationship between the price of apples (X) and per kilo sales (Y) is linear. This gives the theoretical regression model:

$$Sales = \beta_0 + \beta Price + \varepsilon$$

The data in Table 10.4 was collected over a period of 10 days. Each day shows how many kilos of apples were sold and at what price (in kronor). The SPSS file is called Apple sales.

Using OLS estimation, the estimated model for observation i is:

$$Sales_i = 50.3 - 1.55 Price_i + e_i$$

The predicted value of $\widehat{Y}$ (sales), for observation i, is computed by the following equation:

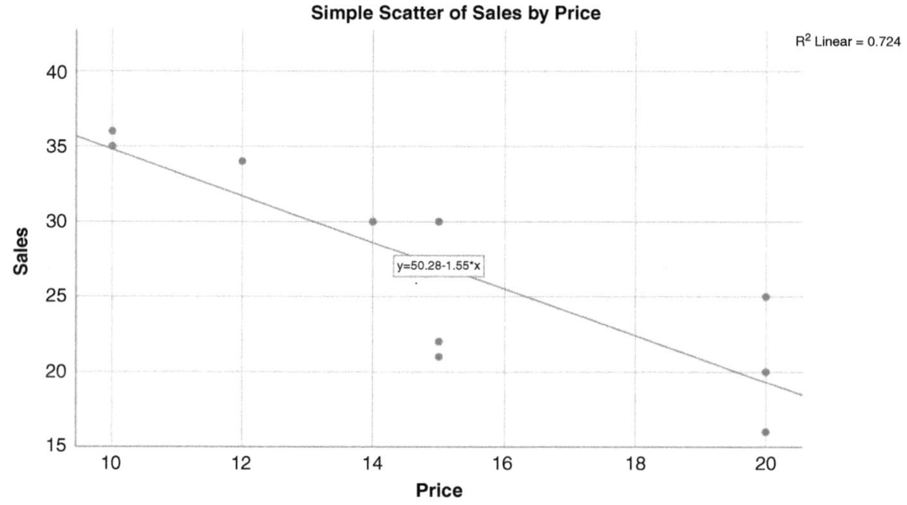

Fig. 10.9 Regression plot for apple sales

$$\widehat{Sales_i} = 50.3 - 1.55 Price_i$$

Note that we include a $\widehat{hat}$ over the parameter/variable that we predict. Assume that the price of apples is 12 kroner today. An increase of 2 kronor will reduce sales by −1.55*2 = −3.1 kilos. Figure 10.9 shows the graph of the regression line. With a negative beta coefficient for the *X* variable, the slope is negative.

The constant (*Y*-intercept) suggests that when the price is zero, sales will be 50.3 kilos. This is an excellent example of what we mentioned earlier about the absurd values that the intercept can have. Our "theory of apple pricing" probably holds above a certain minimum price and below some maximum price. This is the same for many theories. They apply within a range of values. However, sometimes the constant is important from a theoretical perspective so do not necessarily ignore it, and it must always be included in OLS regression. ◄

Example 10.2

In this example, the constant is very important. Assume that we estimate a regression model for the gender pay gap (*Y*) and seniority (*X*), for men and women employed at a consultancy. The regression equation for the predicted value of $\widehat{Y}$ for observation *i* is:

$$\widehat{Y}_i = 10 + 5.5X_i$$

Y is measured as the difference in pay for the sum of all women minus the sum for all men for each year of seniority. In this example, the constant means that in year zero, when a person is hired, women get, on average, 10 thousand kronor per year more than men. Both groups increase their earnings by 5500 kronor for each year of employment. ◀

10.4 The *T*-test

Imagine that management at a hotel decides to increase service in an effort to increase revenue. They implement a large (expensive) project that includes substantial training of all employees. After implementation, the managers see only small changes in revenue. This shows that strategic choices can have serious financial consequences. To better understand the implementation and its effect on revenue, the managers could run regression analysis. With the variables revenue (Y) and implementation costs (X), they can test whether there is a statistically significant relationship between the implementation and revenue. Specifically, they test whether the regression slope coefficient β_1 is significantly different from zero. In OLS regression, they look at the *t-test of the regression coefficient(s)* (we discuss the F-test and multiple regression later in the chapter). This would show whether the implementation (measured as its cost) affects revenue. The null hypothesis for the slope coefficient is:

H_0: $\beta_1 = 0$, which can also be expressed as: $Y_i = \beta_0 + 0 * X_i + \varepsilon_i = \beta_0 + \varepsilon_i$

In other words, the null hypothesis that Y (revenue) is equal to the constant plus random error. This is the same as saying X has no effect on Y. The investment in service has no effect on revenue.

The alternative is that there is a significant relationship between X and Y. There are three possibilities: one two-sided and two one-sided. The two-sided hypothesis would be that revenue may go up or down as a result of an increase in service. Either costs substantially exceed revenues, leading to a drop in revenue. Or, the cost of the service improvement is substantially outwieghed by the resulting revenue increase. This means that from a two-sided perspective, β_1 is significantly different from zero ($\neq$), either higher or lower. The two one-sided alternatives are to hypothesize that β_1 is significantly less than zero, or that β_1 is significantly greater than zero.

The procedure for all three options is the same, except that in the t-tables you refer to the one-sided or two-sided critical values for a given level of significance (α). t-tables will normally be constructed showing a range of degrees of freedom (down the left side of the table); there are one and two-sided alternatives (across the top of the table), for a given level of significance (α). For example, for large samples (over about 100), the critical cutoff value for a two-sided t-test at a significance level of 5% is 1.96. To determine the degrees of freedom for simple regression, the formula is $n - 2$. For multiple regression (discussed later), the formula is $n - k - 1$, where k is the number of independent variables. The test-statistic for t is calculated by dividing the estimated beta coefficient ($\widehat{\beta}$) by the standard error of the beta coefficient $s_{\widehat{\beta}}$:

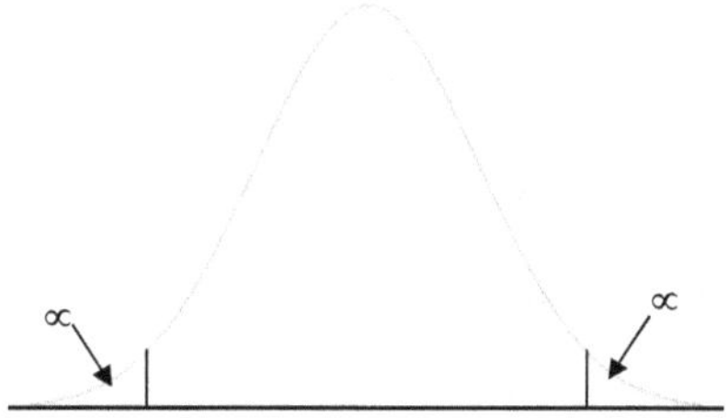

Fig. 10.10 t-distribution: two-sided and one-sided tests

$$t = \frac{\widehat{\beta}}{s_{\widehat{\beta}}}$$

The Two-Sided *t*-test: Positive or Negative
H_0: $\beta_1 = 0$, and H_1: $\beta_1 \neq 0$. H_0 is rejected if $|t| > t_\alpha$. We then say that β_1 is not equal to zero. The sign of β_1 is irrelevant because it can be either positive or negative.

The One-sided *t*-test: Positive
H_0: $\beta_1 \leq 0$, and H_1: $\beta_1 > 0$. H_0 is rejected if $|t| > t_\alpha$ for a one-sided test and the sign of β_1 is positive. Look at the *t*-table, but this time you look at the one-sided alternative for a given level of significance (α). For a large sample (over 100), the critical cutoff value for a one-sided *t*-test at a significance level of 5% is 1.645.

The One-sided *t*-test: Negative
H_0: $\beta_1 \geq 0$, and H_1: $\beta_1 < 0$. H_0 is rejected if $|t| > t_\alpha$ for a one-sided test and the sign of β_1 is negative. Exactly the same as for the one-sided (positive) *t*-test above, look at the *t*-table, for the one-sided alternative for a given level of significance (α). The *t*-table does not distinguish between positive and negative, so for a large sample (over 100), the critical cutoff value for a one-sided *t*-test at a significance level of 5% is 1.645.

Visually, the *t*-distribution can be expressed as in Fig. 10.10. At each end (tail) of the distribution, there is an area designated as alpha (α), which is the significance level of the test. Two-sided tests use both tails, whereas one-sided tests use one tail. You decide the level of significance and whether it is a one-sided or two-sided test. Then, based on the degrees of freedom (ν), you find the critical cutoff value. For each degree of freedom, the distribution curve changes slightly.

Example 10.3

Picking up with the apple sales again, we assume that when prices rise, sales drop. The hypotheses are:

H_0: $\beta_1 \geq 0$, and H_1: $\beta_1 < 0$

We use a 5% significance level. Our OLS regression estimates gave these values:

$\beta_0 = 50.28$ and $\beta_1 = -1.55$

We can calculate the test-statistic for t using the formula:

$$t = \frac{\widehat{\beta}}{s_{\widehat{\beta}}} = \frac{-1.55}{0.338} = -4.58$$

This is confirmed by looking at the coefficients output generated in SPSS:

If we want to compare the test-statistic (−4.579) against the critical cutoff value in the t-tables, we take $n - 2 = 10–2 = 8$ degrees of freedom. This is a one-sided test because we hypothesized a negative relationship, and we decided a 5% significance level. The cutoff value from the tables is: 1.860. Since the absolute value of $|-4.579| > 1.860$, we reject the null hypothesis. We conclude that price has a negative relationship with sales of apples. We can also draw this conclusion by simply looking at the p-value (significance probablility). Since $0.002 < 0.05$, we reject the null hypothesis. ◀

10.5 Multiple Regression Analysis

We can extend simple regression, where there is one independent variable, to *multiple regression*, where there are two or more independent variables. We began the chapter suggesting our theory that revenue is a function of occupancy, price, advertising, competition, and service. We can formalize the hypotheses as:

H1: Occupancy has a positive effect on revenue.
H2: Price has a negative effect on revenue.
H3: Advertising has a positive effect on revenue.
H4: Competition has a negative effect on revenue.
H5: Service has a positive effect on revenue.

The model is:

We assume that the relationships are linear, which can be expressed in the following theoretical regression equation:

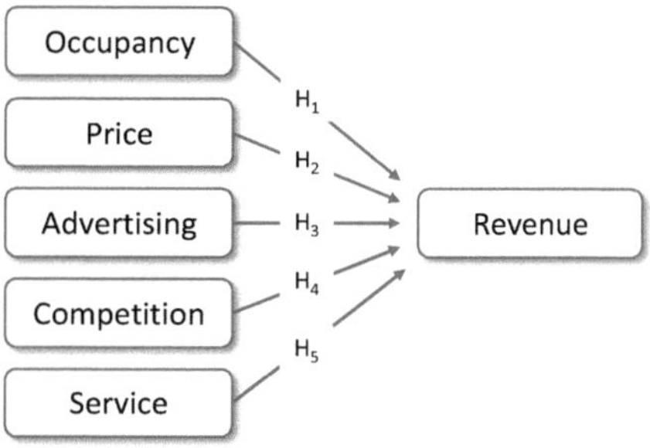

Fig. 10.11 Regression model for hotel revenue

$$Revenue = \beta_0 + \beta_1 Occupancy + \beta_2 Price + \beta_3 Advertising + \beta_4 Competition + \beta_5 Service + \varepsilon$$

The general model for multiple regression can be expressed as:

$$Y_i = \beta_0 + \beta_1 X_{1i} + \beta_2 X_{2i} + \beta_3 X_{3i} + \ldots + \beta_k X_{ki} + \varepsilon_i$$

Where Y is the dependent variable and the Xs are the k independent variables. ε_i is the error term and the i is the ith observation. β_0 is the constant (Y-intercept) and β_i ($i = 1, 2, 3, \ldots, k$) are the regression parameters.

Multiple regression is estimated in the same way as simple regression, with ordinary least squares (OLS) estimation. To run the multiple regression analysis in SPSS, choose: *Analyze > Regression > Linear. Put Revenue into the Dependent box and Occupancy, Price, Advertising, Competition, and Service into the Independent (s) box* (see Fig. 10.11).

10.6 Explained Variance

The first table in the SPSS regression output is the Model Summary (see Table 10.5). With the R^2, you can answer the question, how much of the variance in Y does the regression equation explain?

Total variance can be broken down into two main parts:

$$Total\ variance = explained\ variance + unexplained\ variance.$$

Explained variance is calculated:

$$TSS\ (total\ sum\ of\ squares) = RSS\ (regression\ sum\ of\ squares) + ESS\ (error\ sum\ of\ squares)$$

The fraction $\frac{RSS}{TSS}$ is called the *explained variance*. An alternative name is the *coefficient of determination*, which is abbreviated as R^2. Therefore, we can write:

Table 10.5 Sales and price coefficients

Coefficients[a]						
		Unstandardized coefficients		Standardized coefficients		
Model		B	Std. Error	Beta	t	Sig.
1	(Constant)	50.283	5.255		9.569	0.000
	Price	−1.549	0.338	−0.851	−4.579	0.002

[a]Dependent Variable: Sales

$$R^2 = \frac{RSS}{TSS}, \text{or } R^2 = 1 - \frac{ESS}{TSS}$$

R^2 (expressed as r-square) has a value between 0 and 1, with values closer to 1 indicating higher explained variance. For example, when R^2 is above 0.5, it means that more than half of the variance in Y is explained by the regression equation and the rest is error. When R^2 is low, there are probably other important explanatory independent variables missing from the equation. In this situation, we could search for other independent variables that increase the explanatory power.

Early in this chapter we ran a simple regression between occupancy and revenue. The R^2 was only 11.2%. That means that occupancy only explains 11.2% of the variance in revenue. In this situation, it would be good to consider adding independent variables. In our new model (Fig. 10.12), we include four more independent variables (price, advertising, competition, and service). The Model Summary output in Table 10.6 shows that the explained variance (R^2) has increased to 42.5%. While this is a vast improvement, there must be other variables we could add to substantially improve the model. Of course, we would have to have them in our dataset.

R^2 is derived through the following equation:

$$R^2 = \frac{ESS}{TSS} = \frac{17,339.917}{40,752.990} = 0.425 = 42.5\%$$

The RSS and TSS values come from the ANOVA table in Table 10.6 (below). They are:

$RSS = 17{,}339.917$
$ESS = 23{,}413.073$
$TSS = 40{,}752.990$

An important characteristic of R^2 is that it always increases with the addition of independent variables. The more independent variables you include, the higher R^2 will be, even when the additional variables are irrelevant. This may result in an overly optimistic evaluation of a regression equation when R^2 is used uncritically as a quality criterion.

When comparing two regression models with the same dependent variable, but with a different number of explanatory variables, you should not automatically select

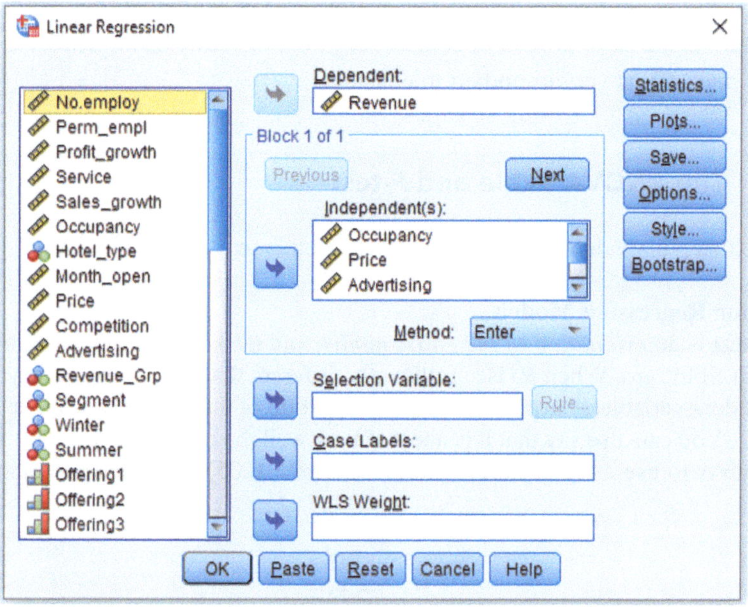

Fig. 10.12 Regression dialogue box modedata

Table 10.6 Model summary for multiple regression

Model summary				
Model	R	R square	Adjusted R square	Std. error of the estimate
1	0.652[a]	0.425	0.395	15.782

[a]Predictors: (Constant), Service, Occupancy, Competition, Advertising, Price

the model with the highest R^2. Always consider the *parsimony criterion*, which means that simplicity in a regression model (or any model) is a virtue. Increasing the number of independent variables to gain tiny increases in explained variance may not be worth the trouble of measuring and including the variables. Fewer explanatory variables often leads to a more general model, which may be preferred. Since R^2 does not capture the element of parsimony, there is the *R-square adjusted* (expressed as $\overline{R}^2$ or R^2_{adj}). As we mentioned earlier, it takes into account the degrees of freedom in the model (*n-k-1*), thus adjusting for additional independent variables. The equation is:

$$R^2_{adj} = 1 - \frac{\frac{ESS}{(n-k-1)}}{\frac{TSS}{(n-1)}} = 1 - \frac{\frac{23,413.073}{(100-5-1)}}{\frac{40,752.990}{(100-1)}} = 39.5\%$$

When the sample size is very large, the difference between R^2 and R^2_{adj} is very small, so it is not so important to consider R^2_{adj}. On the other hand, when the sample size is small, $R^2_{adj}$is very important to consider.

10.7 The ANOVA Table and *F*-test

The next step in evaluating a regression equation is to examine the *F*-test.

F-test for Regression Models

The *F*-test is an *evaluation of the entire model*, and tests whether R^2 is significantly different from zero. When R^2 is significantly different than zero, you can say that the independent variable(s) explain a significant portion of the variance in the dependent variable. You can also say that it is a test of how well the model fits the data. The rule of thumb is to use a significance level of 5% ($\alpha = 0.05$). The general equation for F is:

$$F = \frac{\frac{RSS}{k}}{\frac{ESS}{(n-k-1)}}$$

This is the hypothesis test:

H_0: $\beta_1 = \beta_2 = \ldots \beta_K = 0$
H_1: At least one $\beta_i \neq 0$

The ANOVA output from the *Hotel* data with five independent variables is (Table 10.7):

Substituting the values, we get the equation:

$$F = \frac{\frac{RSS}{k}}{\frac{ESS}{(n-k-1)}} = \frac{\frac{17,339.917}{5}}{\frac{23,412.073}{(100-5-1)}} = 13.92$$

The null hypothesis is rejected if the test-statistic for F is larger than the critical cutoff value from the F-tables. The degrees of freedom can be calculated with the formula: $F_{\alpha,\, k\, -\, 1,\, n\, -\, k\, -\, 1}$

$\alpha = 5\%$, which is the significance level we choose. $K =$ the number of variables we are estimating, and $n = 100$. 5–1 = 4 (the numerator), and 100–5 – 1 = 94 (the denominator). Then, refer to the F-table for α at 5%. The critical cutoff value for $F_{\alpha\, =\, 5\%,\, 4,\, 94}$ is 2.33. Since 13.92 > 2.33, we reject the null hypothesis. The R^2 of 42.5% is significantly different from zero.

Of course, with the advent of computers and their ability to calculate the significance probability (p-value), the tables are somewhat redundant. Since the Sig. = 0.000, and the cutoff is 0.05 (5% significance level), you simply see that $0.00 < 0.05$, and conclude that the F-statistic and thus R^2 are significantly different from zero.

Table 10.7 Hotel multiple regression ANOVA

ANOVA[a]						
Model		Sum of squares	df	Mean square	F	Sig.
1	RSS regression	17,339.917	5	3467.983	13.923	.000[b]
	ESS residual	23,413.073	94	249.075		
	TSS Total	40,752.990	99			

[a]Dependent variable: Revenue
[b]Predictors: (Constant), Service, Occupancy, Competition, Advertising, Price

T-test for Regression Coefficients

The next step is to test whether each independent variable has a significant effect on the dependent variable. It is the same procedure for simple regression. In our example, we want to test whether the occupancy rate, price, advertising, competition, and service have significant effects on hotel revenue. For the *t*-tables, we use the formula, *n*-*k*-1, where *n* is the sample size and *k* is the number of variables. This is the theoretical model:

$$Y = \beta_0 + \beta_1 X_1 + \beta_2 X_2 + \beta_3 X_3 + \beta_4 X_4 + \beta_5 X_5 + \varepsilon$$

Which can also be expressed as:

$$Revenue_i = \beta_0 + \beta_1 Occupancy_i + \beta_2 \text{Price}_i + \beta_3 \text{Advertising}_i + \beta_4 \text{Competition}_i + \beta_5 \text{Service}_i + \varepsilon_i$$

These are the hypotheses for each independent variable:

H_0: $\beta_1 = 0$ and H_1: $\beta_1 > 0$.
H_0: $\beta_2 = 0$ and H_2: $\beta_2 < 0$.
H_0: $\beta_3 = 0$ and H_3: $\beta_3 > 0$.
H_0: $\beta_4 = 0$ and H_4: $\beta_4 < 0$.
H_0: $\beta_5 = 0$ and H_5: $\beta_5 > 0$.

Note that all the hypotheses are one-sided. We choose the rule of thumb 5% significance level. Table 10.8 shows the SPSS regression output for the coefficients. SPSS, as well as most statistical programs, shows p-values (Sig.) for a two-sided test. To get a one-sided test, you have to divide the Sig. value by 2.

For example, for occupancy, the estimate of the beta coefficient is: $\widehat{\beta}_1 = 0.358, s_{\widehat{\beta}} = 0.116$

This gives a t-value of: $t = \frac{\widehat{\beta}_i}{s_{\widehat{\beta}}} = \frac{0.358}{0.116} = 3.08$

The critical cutoff value from the *t*-tables for $100–5 - 1 = 94$ degrees of freedom, at a signfificance level of 5%, for a one-sided test, is 1.660. $3.08 > 1.660$, so we reject the null hypothesis. We conclude that occupancy has a significant positive effect on revenue. The critical cutoff value is the same for all the coefficients because

Table 10.8 Hotel multiple regression coefficients

Coefficients[a]						
		Unstandardized coefficients		Standardized coefficients		
Model		B	Std. Error	Beta	t	Sig.
1	(Constant)	18.780	13.308		1.411	0.161
	Occupancy	0.358	0.116	0.245	3.080	0.003
	Price	−7.054	1.167	−0.519	−6.044	0.000
	Advertising	3.174	1.472	0.177	2.155	0.034
	Competition	−0.585	1.586	−0.032	−0.369	0.713
	Service	0.279	1.587	0.014	0.176	0.861

[a]Dependent variable: Revenue

Table 10.9 P-value one-sided test

	Sig.	Sig./2	Test	Conclusion
Occupancy	0.003	0.001	0.001 < 0.05	Significant
Price	0.000	0.000	0.000 < 0.05	Significant
Advertising	0.034	0.017	0.017 < 0.05	Significant
Competition	0.713	0.357	0.357 > 0.05	Insignificant
Service	0.861	0.431	0.431 > 0.05	Insignificant

they are all one-sided tests. We can save time looking in the *t*-tables by just making a table of the p-values for a one-sided test (see Table 10.9).

Standardized Beta Coefficients

In Table 10.7, there is a column called standardized beta coefficients. It shows standardized effects of all the beta coefficients so that you can compare them for their relative effect on the dependent variable. They can have values between −1 and + 1, and you compare their absolute values for relative effect size. The standardized beta is highest for price (−0.519), second highest for occupancy (0.245), and lowest for advertising (0.177). *Do not* interpret insignificant beta coefficients, so we do not discuss their standardized beta coefficients.

The *practical conclusions* are that a one-unit change in price will have the greatest impact on revenue, followed by occupancy and advertising. It would be up to management to decide which of the variables are most actionable. Perhaps they do not want to start a price war with the competition, and they cannot directly affect occupancy, so they opt to increase advertising to increase revenue. Then, the question would be whether the added costs of advertising are offset by the additional revenue.

Regression Partial Plots

The partial plots make it possible to visually view the contribution of each independent variable onto the dependent variable. Interpretation of the plots and the regression coefficients must be considered within the context of the entire model. Partial

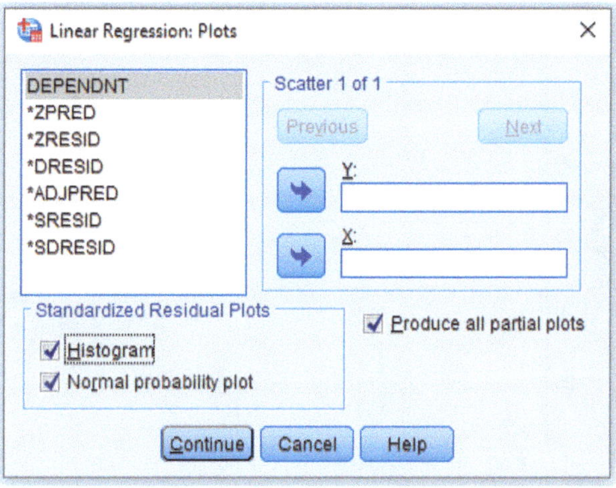

Fig. 10.13 Partial regression plot for occupancy on revenue

plots provide a good way of evaluating the linearity of the relationship between each specific independent variable and the dependent variable. They are also good for identifying multivariate outliers. For example, we highlight one data point on all three plots that may be considered an outlier (observation 83). To generate partial plots, with the dialogue box open for linear regression, *choose Plots, and then tick Produce all partial plots. Click OK* (see Fig. 10.13). *Note that we also requested a Histogram and Normal probability plot (Standardized Residual Plots). These are necessary for later residual analysis.*

Figure 10.14 shows the partial plot for *Price* on *Revenue*. As expected, the slope is negative, meaning that higher prices lead to lower revenue. It is beyond the scope of our discussion. However, it is worth noting that if price was very low, raising the price might increase revenue. This will be related to the price elasticity theory. Nevertheless, within the range of prices we are testing, the relationship is linear and negative.

Observe from the coefficients table, and it is noted on the graph, that the slope of the line is −7.05. The R^2 for the partial regression is 28%, indicating that price accounts for a large portion of the explained variance. This is also observed in the standardized beta coefficients, where price clearly has the largest absolute effect (−0.519).

The occupancy partial plot in Fig. 10.15 has a regression line slope of 0.36, and explained variance of 9.2%. Observation 83 appears to be an outlier.

The advertising partial plot in Fig. 10.16 also shows observation 83 as a possible outlier. Though we will not do it in this analysis, the next step would be to remove observation 83 to see whether the results change in a substantive way. If they do, observation 83 should be left out. One observation should not determine the results of the regression. If there is no substantive change, then it could be kept.

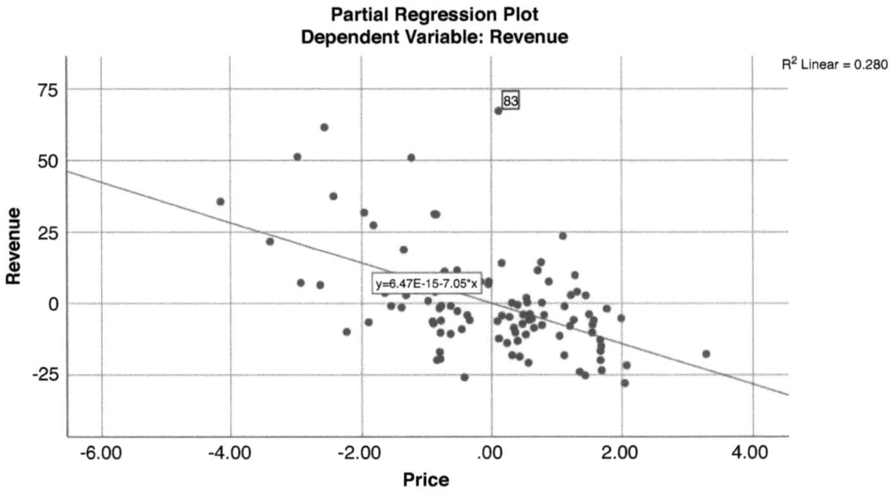

Fig. 10.14 Partial regression plot for Price on Revenue

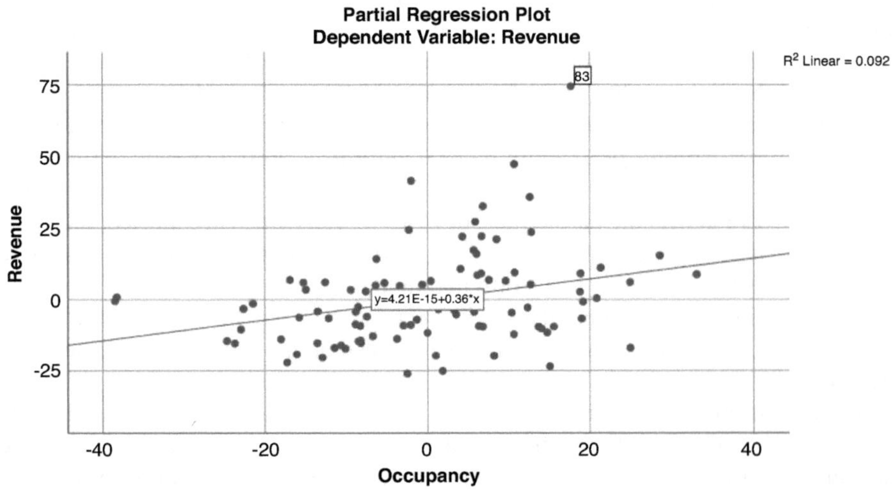

Fig. 10.15 Partial regression plot for Occupancy on Revenue

10.8 Too Many or Too Few Independent Variables

Choosing how many independent variables to include in a regression is a trade-off between parsimony and explanatory power. In our current example with the *Hotel* data, competition and service are not significantly related to revenue, so they can be removed. On the other hand, there are other variables in the dataset that should perhaps be included.

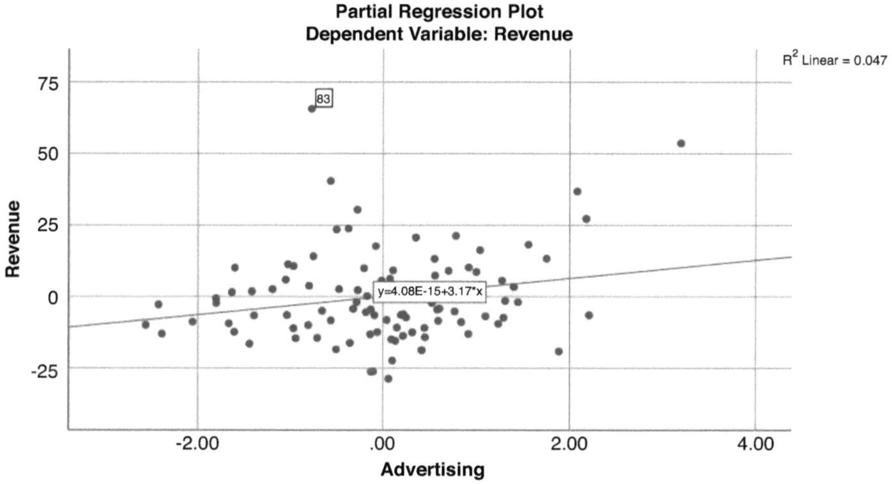

Fig. 10.16 Partial regression plot for Advertising on Revenue

Specification error refers to including irrelevant variables, not including important variables, or choosing the wrong functional form. *Functional form* refers to linear and nonlinear models, which we discuss briefly later. In this section, we restrict the discussion to the consequences of too many or too few independent variables. Having too many independent variables is a luxury problem. You can always remove them. Missing important explanatory variables is a more serious problem. Choosing independent variables is challenging when:

- You do not know which variables are relevant.
- You have limited access to variables.
- You are using proxy variables (substitute variables) and you do not know how well they measure what you intend.

Consequence of a *Too Long* Model

Assume that $Y_i = \beta_0 + \beta_1 X_1 + \varepsilon_i$ is the true (optimal) regression model. Then, add an additional independent variable to get the model:$Y_i = \beta_0 + \beta_1 X_1 + \beta_2 X_2 + \varepsilon_i$. If X_1 and X_2 are correlated with each other, the variance in X_1 will be *randomly affected*. The consequence is to reduce the precision in β_1. The same applies to all the beta coefficients in a multiple regression. A too long model *reduces the precision of the beta coefficients*. In general, insignificant or meaningless independent variables should not be kept in a regression. However, sometimes the researcher wants to make the point that a particular variable is insignificant. Imagine a study on the gender pay gap. Finding that gender is not significant would be very meaningful for the results. Keeping it in the regression may reduce the precision of the other beta coefficients. However, demonstarting that it is insignificant would outweigh the loss in precision.

Table 10.10 Model summary with number of employees

Model summary				
Model	R	R square	Adjusted R square	Std. error of the estimate
1	0.935[a]	0.874	0.869	7.353

[a]Predictors: (Constant), No.employ, Advertising, Occupancy, Price

Table 10.11 ANOVA with number of employees

ANOVA[a]						
Model		Sum of squares	df	Mean square	F	Sig.
1	Regression	35,616.130	4	8904.032	164.669	0.000[b]
	Residual	5136.860	95	54.072		
	Total	40,752.990	99			

[a]Dependent variable: Revenue
[b]Predictors: (Constant), No.employ. Advertising. Occupancy, Price

Consequence of a *Too Short* Model

Assume that $Y_i = \beta_0 + \beta_1 X_1 + \beta_2 X_2 + \varepsilon_i$ is the true regression model. Assume that X_2 is dropped from the equation. The resulting model is: $Y_i = \beta_0 + \beta_1 X_1 + \varepsilon_i$. If X_1 and X_2 are correlated, then the *estimated beta coefficient* $\widehat{\beta}_1$ *will be systematically biased.* This is because the parameter estimates are a function of the correlation between the included and excluded independent variables. The *estimated error variance will also be miss-specified,* which leads to *flawed confidence intervals.* In sum, missing important independent variable(s) causes a *systematic bias in the parameter estimates,* which is more serious than just losing precision.

For an example of a too short model, in our current hotel example, add *number of employees* to the independent variables, and remove competition and service. The results are shown in Tables 10.10, 10.11, and 10.12.

Note the dramatic increase in explained variance (R^2_{adj}), from 39.5% (see Table 10.5) to 86.9% in Table 10.10. Clearly, number of employees is an extremely important variable for explaining revenue.

In Table 10.11, the significance probablility (p-value) is below 0.05, indicating that the R^2 is singificantly different from zero. The equation fits the data well.

In Table 10.12, we see that the variables price and number of employees significantly influence revenue, whereas occupancy and advertising have become insignificant. This means that the influence of the number of employees variable is so strong that the effects of occupancy and advertising disappear. We re-estimate the model without occupancy and advertising.

The R^2 for the final model is 87.3%, with a corresponding F-statistics of 333.062 (df. 2, 97), and a p-value of 0.000. We conclude that the model fits the data well. The coefficients table is shown in Table 10.13.

According to the unstandardized betas, price has a negative effect on revenue, whereas the number of employees has a positive effect. The standardized betas show that, by absolute value, the number of employees has a much larger effect ($0.846 > |-0.160|$). A one-unit increase in price will reduce revenue by

Table 10.12 Coefficients with number of employees

Coefficients[a]

		Unstandardized coefficients		Standardized coefficients		
Model		B	Std. Error	Beta	t	Sig.
1	(Constant)	13.090	4.665		2.806	0.006
	Occupancy	0.004	0.058	0.002	0.063	0.950
	Price	−2.179	0.569	0.160	−3.831	0.000
	Advertising	−0.619	0.695	−0.035	−0.891	0.375
	No.employ	0.570	0.031	0.857	18.405	0.000

[a]Dependent variable: Revenue

Table 10.13 Final model

Coefficients[a]

		Unstandardized coefficients		Standardized coefficients		
Model		B	Std. Error	Beta	t	Sig.
1	(Constant)	11.360	2.933		3.874	0.000
	Price	−2.173	0.564	−0.160	−3.850	0.000
	No.employ	0.563	0.028	0.846	20.380	0.000

[a]Dependent variable: Revenue

2.172 units. A one-unit increase in employees will have a 0.563 unit increase on revenue. From a pragmatic perspective, increasing price means fewer guests come and revenue drops. Increasing the number of employees probably improves service in some way that attracts more guests, so revenue increases. These effects apply to this range of data. At more extreme levels, the effects probably become nonlinear. For example, radically increasing the number of employees will raise costs to a point where revenue falls.

10.9 Regression with Dummy Variables

Several times throughout the book we have discussed nominal level variables, that is, categorical variables with no order. *Dummy variables* are categorical binary variables usually coded with the values 0 and 1. In ordinary least squares regression, they can be included as independent variables. They cannot be a dependent variable. There are special forms of regression for handling categorical dependent variables, for example, logistic regression. However, this is beyond the scope of this textbook.

A requirement for testing group differences with dummy variables in OLS regression is that there must be *k-1* dummy variables for *k* categories. In the *Hotel* data, there is a *Segment* variable that has three categories: vacation, conference, and business. In the next section, we will demonstrate how to recode variables to form dummy variables. Then, we will show how to include them in OLS regression.

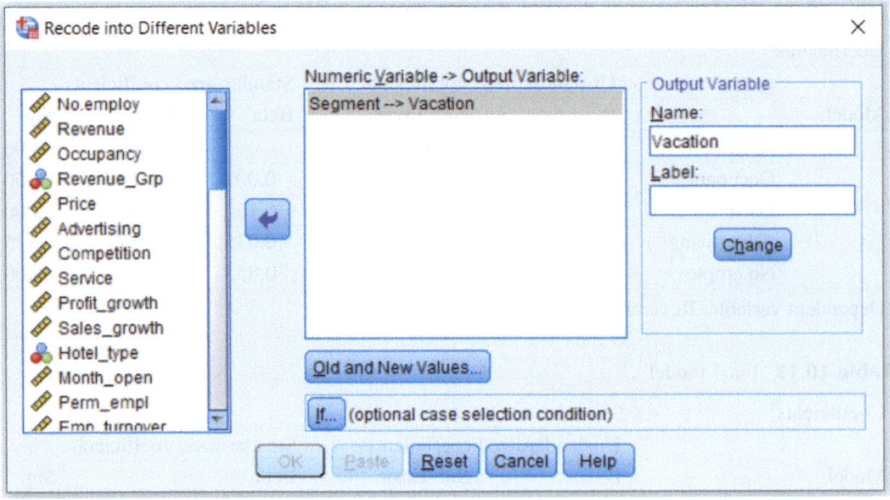

Fig. 10.17 Recode into different variable

Forming Dummy Variables

In this example, we will investigate whether there is a difference in the dependent variable *Revenue*, for the three hotel segments: vacation, conference, and business. We will also include the independent variable, *Occupancy*. Following the *k*-1 dummy rule, we need to form two dummy variables. In the *Hotel* dataset, we have already created them for you, called *Vacation* and *Conference*. Nevertheless, we will show you how to create them.

In SPSS, *choose Transform > Recode into Different Variables. Move the Segment variable into the Numeric Variable position. Name the Output Variable something (we chose Vacation), and move it to the Output Variable position* (see Fig. 10.17).

Click on the Old and New Values button to open the Recode dialogue box. In the Recode dialogue box, insert 1 as the Old Value and 1 as the New Value, and click Add. Continue by putting 2 in the Old Value and 0 in the New Value, then click Add. Finally, put 3 in the Old Value and 0 in the New Value, and click Add (see Fig. 10.18). This means that all values coded 1 in the Segment variable (Vacation) will be coded 1, while the other two categories will be coded 0. Click Continue and OK, to create the new dummy variable, *Vacation*.

Repeat the process to create the second dummy variable for *Conference*. However, this time category 2, which represents conference hotels, is coded 1, and all other categories are coded 0 (see Fig. 10.19).

As a result, there will be two dummy variables where the *Vacation* dummy variable is coded 1 for each vacation hotel and 0 for all other hotels. The *Conference* dummy variable is coded 1 for each conference hotel and 0 for all other hotels. Business hotels are represented in the two dummy variables as the category that is always coded 0.

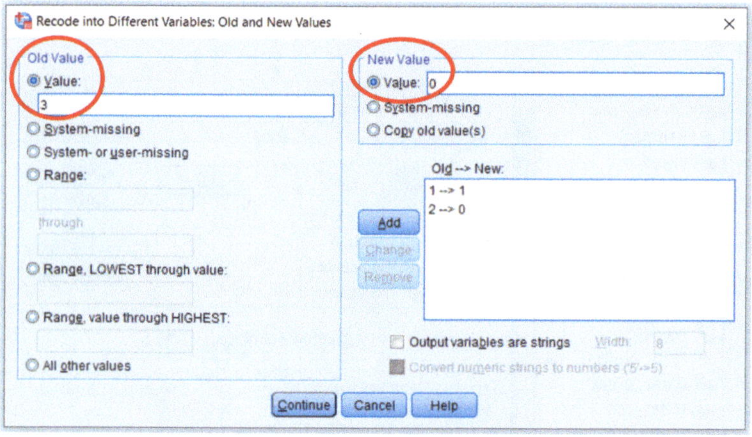

Fig. 10.18 Recode into different variables: old and new values

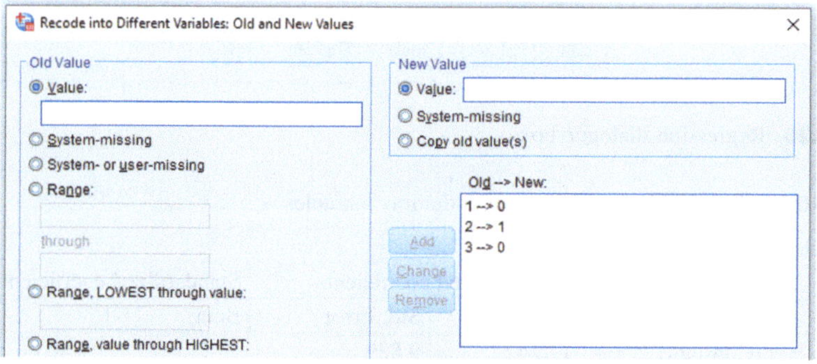

Fig. 10.19 Recode into different variables: old and new values

Figure 10.20 shows the regression dialogue box with *Revenue* as the dependent variable, the two dummy variables *Vacation* and *Conference*, and the continuous independent variable *Occupancy*. The dummy variables are simply treated as additional independent variables. In plain terms, we are testing for segment group differences while controlling for the level of occupancy.

The R^2 for the equation is 26.6%, with an F-statistic of 11.585 and 3, 96 degrees of freedom. The p-value for the F-statistic is 0.000, indicating that the R^2 is significantly greater than zero. Table 10.14 shows the output for the regression coefficients. Applying the rule of thumb cutoff of 0.05, the p-values (Sig.) indicate that all independent variables have a statistically significant relationship with the dependent variable. *Vacation* has a p-value of 0.000, *Conference* has 0.001, and *Occupancy* has 0.043. The easiest way to explain their interpretation is to show a graph.

Figure 10.21 shows the positive slope of the *Occupancy* regression line (0.292), intersecting the Y-axis at 17.74 (rounded off). This is the regression line that

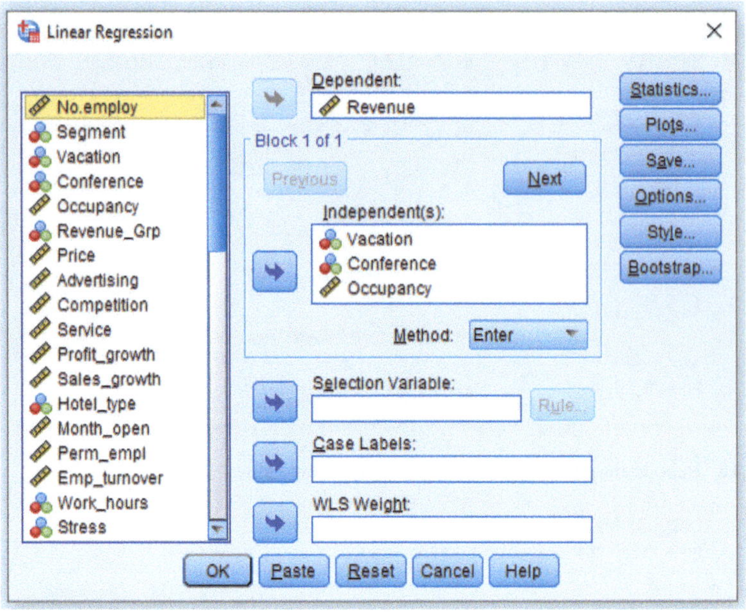

Fig. 10.20 Regression dialogue box

Table 10.14 Regression coefficients with dummy variables

Coefficients[a]						
		Unstandardized coefficients		Standardized coefficients		
Model		B	Std. Error	Beta	t	Sig.
1	(Constant)	17.742	9.829		1.805	0.074
	Vacation	−20.369	4.784	−0.498	−4.258	0.000
	Conference	−16.550	4.671	−0.382	−3.543	0.001
	Occupancy	0.292	0.142	0.199	2.056	0.043

[a]Dependent variable: Revenue

represents business hotels. This is because it is the category that is always zero when combining the *Vacation* and *Conference* dummy variables. To understand the effect of the dummy variables, start by adding the unstandardized beta coefficient for *Conference* (−16.55) to the constant (17.74) to get the *Y*-intercept of 1.19. Then, add the unstandardized beta coefficient for *Vacation* (−20.369) to the constant (17.74) to get the *Y*-intercept of −2.63.

As the graph in Fig. 10.21 indicates, dummy variables shift the regression line up or down from the baseline category, which in this case is business hotels. Business hotels generate more revenue at each level of occupancy than conference hotels, followed by vacation hotels. This example shows the danger of interpreting the intercept. Presumably, none of the hotel types would generate revenue with zero

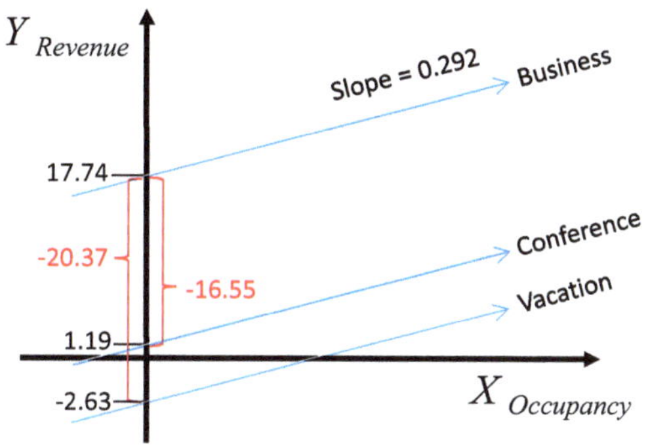

Fig. 10.21 Regression ANCOVA graph

occupancy. Nevertheless, it shows the degree by which revenue increases as occupancy increases for each segment.

To summarize, the continuous independent variable *Occupancy* forms a regression line with a *Y*-intercept at the constant. The dummy variables shift the line up or down by the magnitude of their respective unstandardized beta coefficients.

10.10 Dummy Regression: An Alternative Analysis of Covariance ANCOVA

In Chap. 9, we explained the independent samples t-test as a way of testing mean differences between two independent samples. The independent samples *t*-test is directly related to the *t*-test of the regression coefficients. In Chap. 9, we tested for mean differences of *Revenue* for *Hotel type*: boutique coded 0 and chain coded 1. The mean revenue for each hotel type is shown in Table 10.15.

The difference is statistically significant with a *t*-statistic of −2.276 and corresponding p-value of 0.025. If we instead run an OLS regression with *Revenue* as the dependent variable and *Hotel type* as the independent variable, we get the coefficients table shown in Table 10.16.

Note that the absolute value of the *t*-statistic is the same (2.276) as well as the p-value (0.025). The conclusion from the OLS regression based on the test of the dummy variable *Hotel type* is the same. There is a significant difference between group means for the dependent variable, *Revenue*. The mean revenue for the category coded 0 (boutique) is equal to the constant (15.450). The mean revenue for the category coded 1 is the constant + the unstandardized beta coefficient.

$$15.450 + 9.233 = 24.683$$

When simply testing for group differences on a continuous dependent variable, it is simpler to use an independent samples t-test, or if there are more than two groups, a one-way ANOVA with post hoc tests (also explained in Chap. 9). However, for ANCOVA (analysis of covariance) we suggest using OLS regression.

Table 10.15 Mean revenue differences for hotel type

Group Statistics

	Hotel_type	N	Mean	Std. Deviation	Std. Error Mean
Revenue	Boutique	40	15.45	16.934	2.678
	Chain	60	24.68	21.598	2.788

Table 10.16 Mean revenue differences for hotel type in regression

Coefficients[a]

		Unstandardized coefficients		Standardized coefficients		
Model		B	Std. Error	Beta	t	Sig.
1	(Constant)	15.450	3.142		4.917	0.000
	Hotel_type	9.233	4.057	0.224	2.276	0.025

[a]Dependent variable: Revenue

ANCOVA is a combination of ANOVA and regression, falling under the heading of general linear models. In SPSS, under *Analyze*, you will find the *General Linear Model* category, with several subcategories. In simple terms, you test for group mean differences in a continuous dependent variable, while at the same time testing for covariance in additional continuous independent variables. We just did this in the regression with *Revenue* as the dependent variable, *Occupancy* as a continuous independent variable, and the two dummy variables Conference and Vacation to test for group mean differences across the three segments of vacation, conference, and business hotels.

Table 10.17 shows the group means for *Revenue* by *Segment.*

Table 10.18 shows the coefficients for an OLS regression with *Revenue* as the dependent variable, and *Vacation* and *Conference* as the independent variables. The constant (36.654) is the mean for the group always coded 0, which in this case is business hotels (36.65). To get the means for the other two categories, simply add their respective unstandardized coefficients to the constant:

$$\textit{Vacation} = 36.654 + (-24.201) = 12.45$$

$$\textit{Conference} = 36.654 + (-17.185) = 19.47$$

The *t*-statistic and corresponding p-value indicate whether there is a significant difference between the category represented by the dummy variable and the category

Table 10.17 Means for revenue by segment

Report			
Revenue			
Segment	Mean	N	Std. Deviation
Vacation	12.45	42	7.902
Conference	19.47	32	14.037
Business	36.65	26	30.047
Total	20.99	100	20.289

Table 10.18 Regression coefficients for segments

Coefficients[a]						
		Unstandardized coefficients		Standardized coefficients		
Model		B	Std. Error	Beta	t	Sig.
1	(Constant)	36.654	3.519		10.415	0.000
	Vacation	−24.201	4.478	−0.592	−5.404	0.000
	Conference	−17.185	4.738	−0.397	−3.627	0.000

[a]Dependent variable: Revenue

represented by the constant. With a p-value of 0.000, the mean for Vacation is significantly different from the mean for Business. With a p-value of 0.000 the mean for Conference is significantly different from the mean for Business.

Since there are no post hoc tests in OLS regression, the only way to test for a significant difference between the pairing of vacation and conference is to drop one of them from the regression and replace it with the third category dummy, which in this case is business. Table 10.19 shows the OLS regression coefficients.

The p-value of 0.099 for conference indicates that there is no statistically significant difference between mean revenue for vacation and conference hotels.

The Advantage

While there is the hassle of testing the final category for significant mean differences, there is an advantage with respect to analysis of covariance ANCOVA. In our example including *Occupancy* as an additional independent variable, we are testing the covariance of occupancy on revenue when controlling for segment through the two dummy variables. We can include several independent variables, both continuous and dichotomous dummies.

10.11 The Classic Assumptions of Multiple Regression

There are seven assumptions upon which OLS regression is based. OLS regression is quite robust against deviations from the assumptions. Nevertheless, they should be considered. In practice, we address some of the assumptions as we prepare the data

Table 10.19 Regression coefficients replacing vacation with business

Coefficients[a]						
		Unstandardized coefficients		Standardized coefficients		
Model		B	Std. Error	Beta	t	Sig.
1	(Constant)	12.452	2.769		4.497	0.000
	Conference	7.016	4.211	0.162	1.666	0.099
	Business	24.201	4.478	0.526	5.404	0.000

[a]Dependent variable: Revenue

for regression analysis. Then, when we do the regression analysis we follow it up with a residual analysis. Often, it is an iterative process until we have a regression model that satisfies the assumptions and represents the data.

For testing the assumptions, we use the final *Hotel* data regression model with revenue as the dependent variable, and price and number of employees as the independent variables.

(1) The Regression Model Is Linear in the Coefficients, Is Correctly Specified, and Has an Additive Error Term

So long as we have done a thorough job of specifying the model, based on theory and logic, we assume this assumption to be satisfied. There are a few important things to know. The equations do not have to be linear in the variables, only in the parameters. This allows for using nonlinear variables, like quadratic functions, as variables. We discussed the problems associated with too short and too long models. We specify the best model we can under the circumstances we are presented with. Finally, all regressions must include an error term. So long as you do not specify anything in the software other than to include the error term (it is usually default), this assumption is satisfied.

We can get an indication of whether there may be problems with this assumption by examining the partial plots and a scatterplot of the *standardized predicted values* (*X*-axis) on the standardized residuals (*Y*-axis). For the partial plots, *choose analyze > regression > Linear. Put revenue in the dependent box and choose price and number of employees for the independent variables. Open Plots, and choose Produce all partial plots. For use in subsequent analysis, also choose Histogram and Normal probability plot* (refer back to Fig. 10.13). *Click continue. Then, click Save, and choose Standardized predicted values and Standardized residuals* (see Fig. 10.22). *Click Continue and OK.*

The partial plots are shown in Figs. 10.23 and 10.24. We have taken the liberty of identifying potential outliers. We will not do anything about outliers in the current example. However, we should test the results without the outliers to see if the estimates substantively change. Aguinis et al. (2013) offer an excellent discussion of how to address outliers.

Despite the outliers, the number of employees has a strong linear relationship with revenue. We do not see any unusual patters or a tendency toward being curved.

Fig. 10.22 Regression save

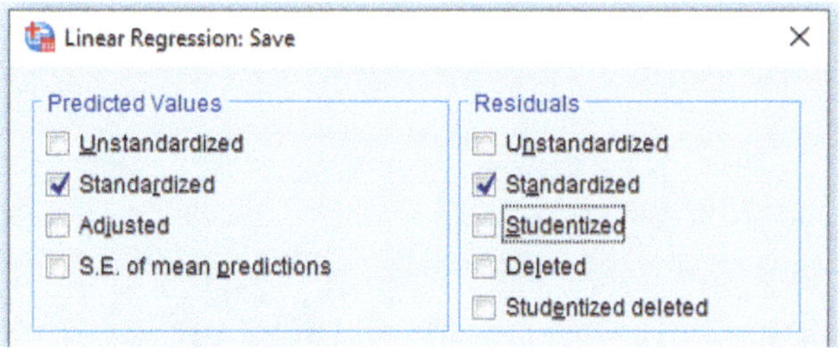

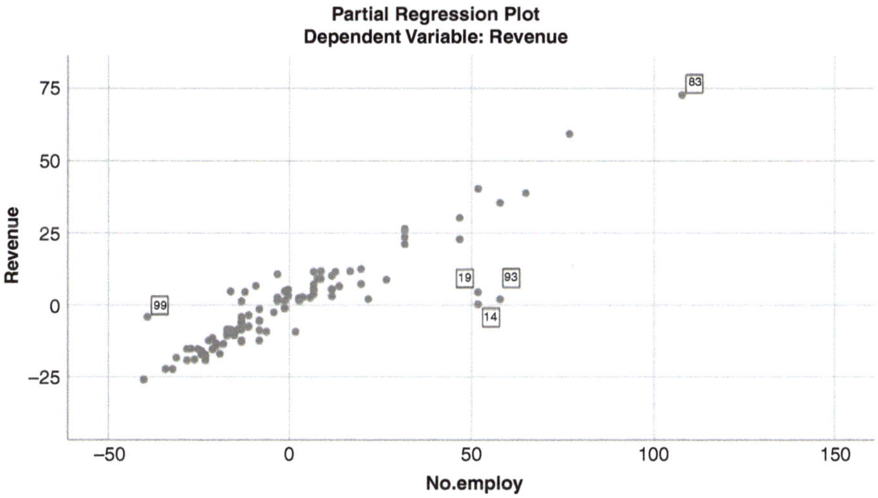

Fig. 10.23 Partial plot for number of employees

Price has less of a clear slope. However, it appears linear, especially once outliers are removed.

To estimate the residual plot, *choose Graphs > Chart Builder > Scatter/Dot. Drag and drop a simple scatter into the Gallery. Drag the new variables we saved to their appropriate axes. Choose OK* (see Fig. 10.25).

The output is shown in Fig. 10.26. Again, we have taken the liberty to highlight outliers. The data, in spite of the outliers, is somewhat clumped together. However, it appears random and there is no nonlinear pattern. Our conclusion is that assumption 1 is met.

(2) The Error Term Has a Zero Population Mean

This assumption cannot be tested. It is assumed to be satisfied so long as the constant (β_0) has been included in the equation.

(3) The Independent Variables Are Uncorrelated with the Error Term

This happens, for example, when something that is unaccounted for in the independent variables, meaning it is in the error term, is correlated with one or more of the

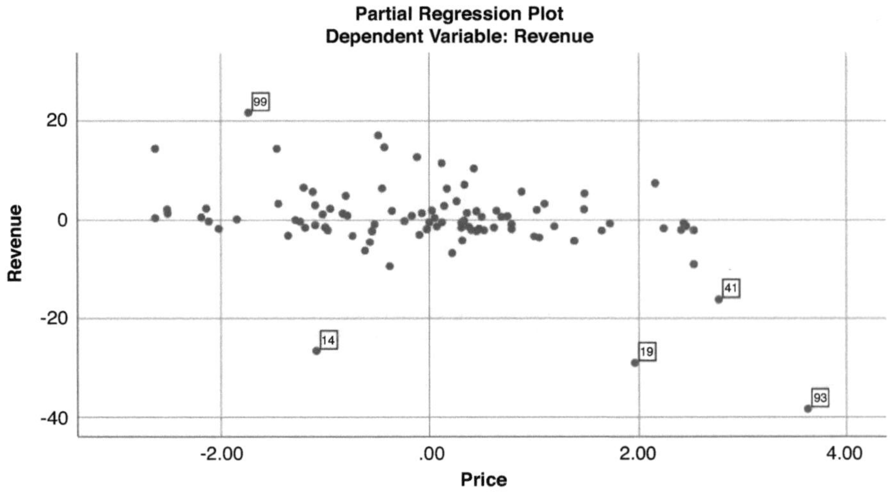

Fig. 10.24 Partial plot for price

independent variables. In recent years, this is referred to as the *endogeneity problem*. In our regression example, we modeled price and number of employees as independent variables onto revenue as a dependent variable. We have assumed a linear relationship. What if, which might easily be the case, the hotels raise their prices and number of employees in high season?

Season affects revenue, price, and staffing. If the seasonal changes are substantial to the extent that they would be statistically significant, then because they are in the error term, the independent variables are correlated with the error term. It also means that the unaccounted effect simultaneously influences the dependent and independent variables. This assumption highlights the importance of proper model specification.

(4) Uncorrelated Error Terms (no Serial Correlation)

This assumption is often violated in time series analysis. The residuals from data collected in one time are correlated with the residuals from data collected in another time. To test this assumption, plot any sequencing variables, like time, with the standardized residuals. If patterns emerge, then the assumption is violated. Unless there is a sequencing variable in the data, this assumption is usually satisfied.

(5) The Residuals Have a Constant Variance (no Heteroscedasticity)

To test for heteroscedasticity, plot the standardized residuals against the standardized predicted values (refer back to Fig. 10.26). If there are no distinct patterns and the data is randomly distributed, this assumption is satisfied.

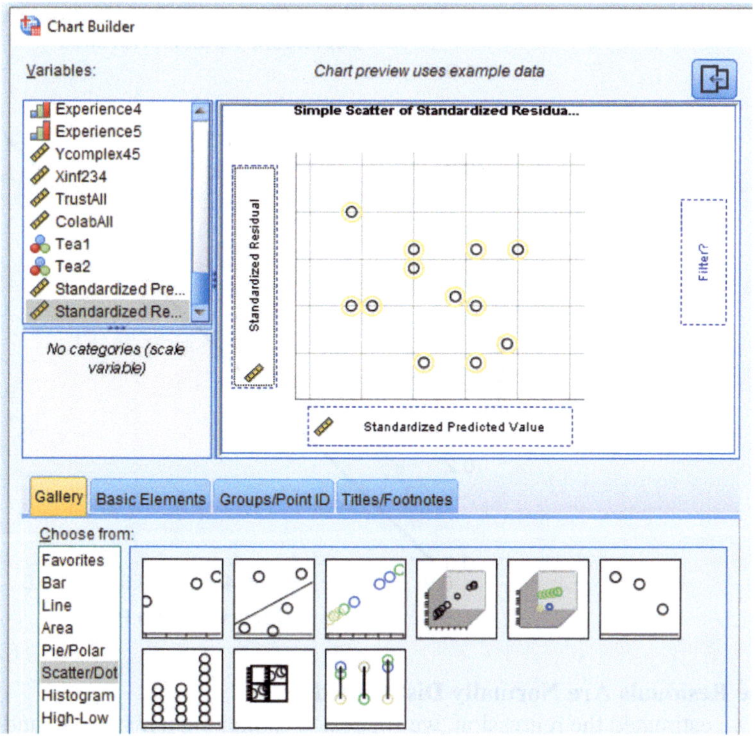

Fig. 10.25 Plotting a residual analysis

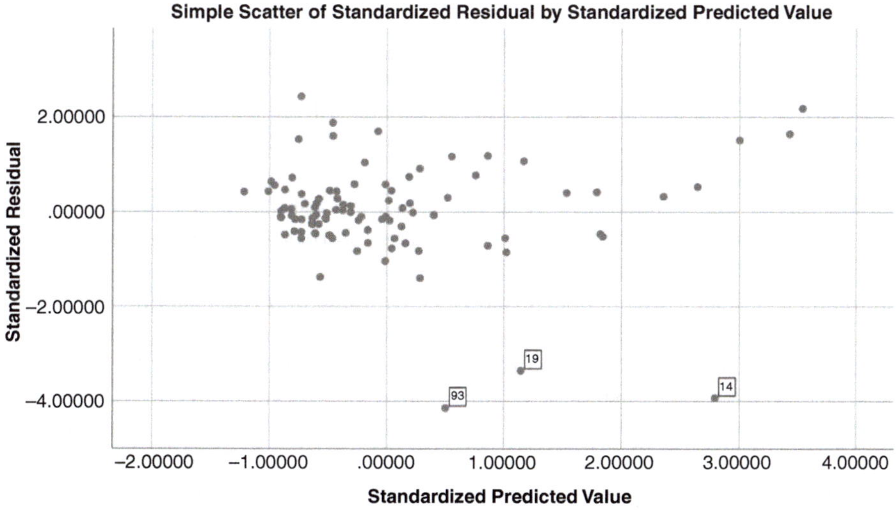

Fig. 10.26 Residual plot

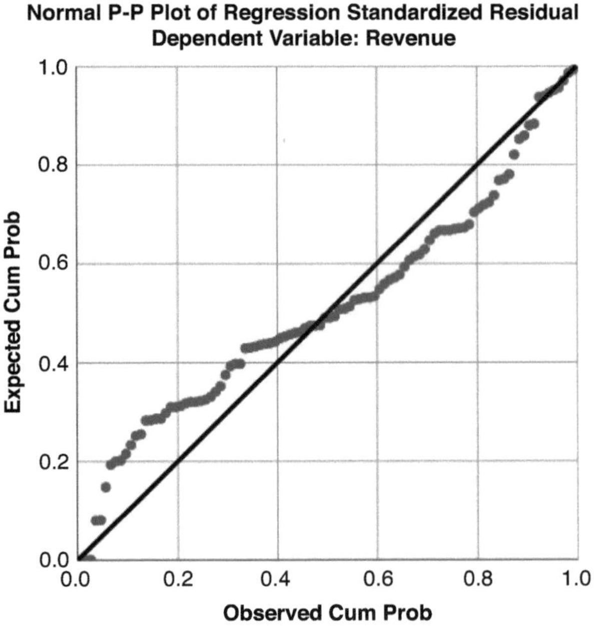

Fig. 10.27 Normal P-P plot

(6) The Residuals Are Normally Distributed

When we estimated the regression, we suggested estimating a histogram and normal probability plots. The normal P-P plot, for a normal distribution, should show the data tightly clumped along the line. In Fig. 10.27, the data shows a slight S-curve along the line.

The histogram in Fig. 10.28 appears peaked and drawn out to the left.

Another way to test for normality is to look at the skewness and kurtosis statistics for the standardized residuals. *Skewness* is how much the data is skewed in either a positive or negative direction. *Kurtosis* is a measure of how peaked or flat the distribution is. Most computer software normalizes these measures so that normal is when they are between -1 and 1. Another way to put it is skewness and kurtosis should be below an absolute value of 1.

We saved the standardized residuals when we estimated the regression (see Fig. 10.22). *Choose Analyze > Descriptive Statistics, Descriptives. Put the standardized residuals into the Variable(s) box. Choose options and untick everything, then tick skewness and kurtosis. Click continue and OK* (see Fig. 10.29).

Table 10.20 shows the results. The skewness of -1.297 suggests that the distribution is slightly skewed to the right. Given the robustness of OLS regression analysis, this skewness, which is slightly above the absolute value of 1, would not overly concern us. The kurtosis is 5.878, which is way above the recommended cutoff of 1. This leads us to question the validity of the regression analysis.

For the sake of simplicity in explaining the regression example, we have not removed the outliers. In a normal analysis, we would have tested the regression

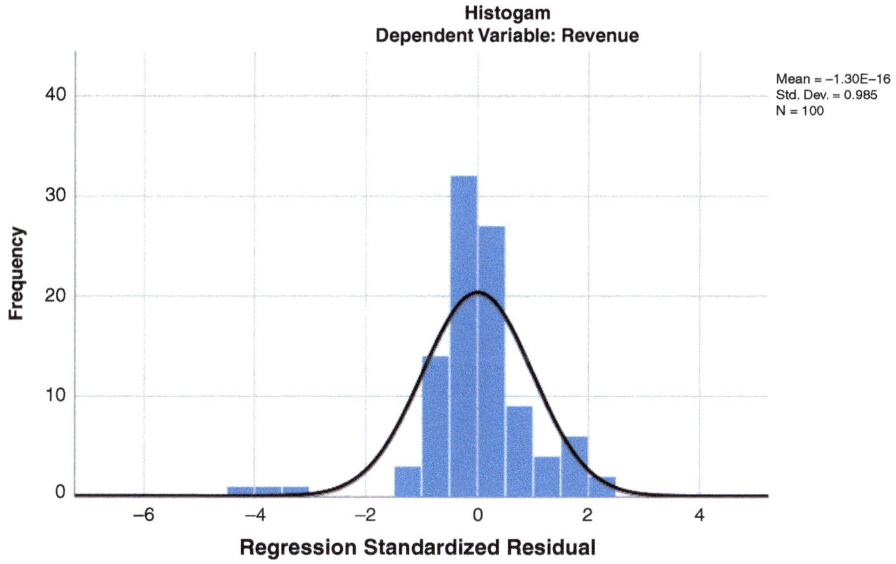

Fig. 10.28 Residual histogram

without the outliers. In fact, by removing four outliers the regression performed much better with respect to the assumptions. Interestingly, it did not change the interpretation. This points to the fact that regression is quite robust even when the assumptions are violated.

(7) No Multicollinearity

No independent variable should be a perfect linear function of another independent variable. That is, there is no perfect multicollinearity. If two independent variables are perfectly correlated, then in a regression, they explain the same thing. There is no point to having both variables in the equation. The generally accepted cutoff level indicating a problem with multicollinearity is a correlation coefficient higher than an absolute value of 0.9. The first test of multicollinearity is to run a correlation matrix between the independent variables. We show the correlation matrix for price and number of employees in Table 10.21. At 0.489, it is well below the accepted cutoff.

If the correlation coefficient was high enough to raise our concerns, then in the regression analysis we could ask for Collinearity diagnostics. *Choose statistics, and tick collinearity diagnostics* (see Fig. 10.30).

They show up to the right of the coefficients table (see Table 10.22).

The generally accepted cutoff indicating a problem with multicollinearity is a VIF number above 10 (Hair, Black, Babin & Anderson, 2014). VIF stands for variance inflation factor. At 1.314, it is well below the cutoff. We conclude that this assumption is satisfied.

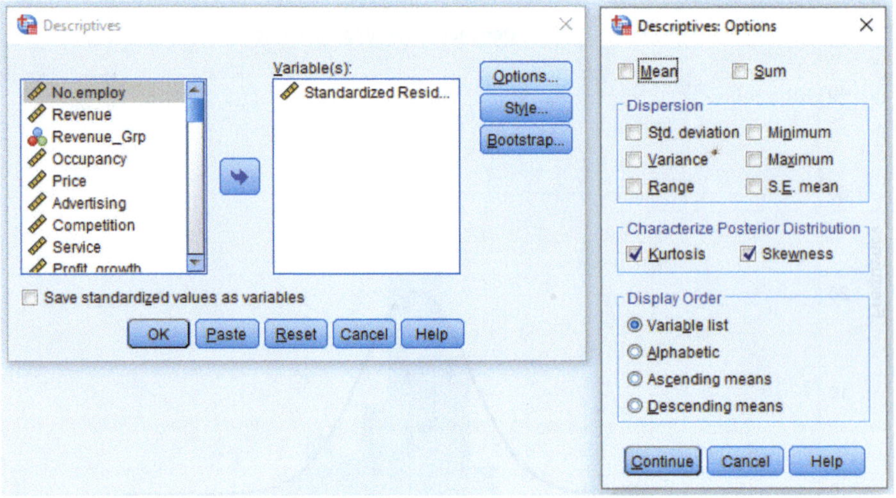

Fig. 10.29 Residual analysis

Table 10.20 Skewness and kurtosis

Descriptive statistics					
	N	Skewness		Kurtosis	
	Statistic	Statistic	Std. error	Statistic	Std. error
Standardized residual	100	−1.297	0.241	5.878	0.478
Valid N (listwise)	100				

Table 10.21 Correlation for multicollinearity

Correlations			
		No.employ	Price
No.employ	Pearson correlation	1	-.489[a]
	Sig. (2-tailed)		0.000
	N	100	100
Price	Pearson correlation	-.489[a]	1
	Sig. (2-tailed)	0.000	
	N	100	100

[a]. Correlation is significant at the 0.01 level (2-tailed)

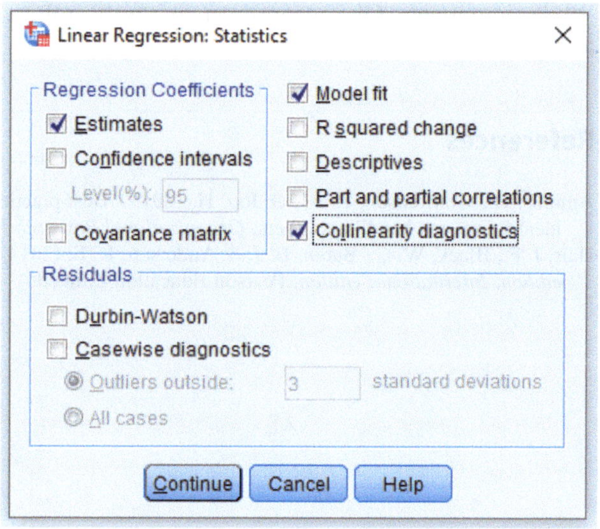

Fig. 10.30 Collinearity diagnostics

Table 10.22 Variance inflation factor

Coefficients[a]

Model		Unstandardized Coefficients B	Unstandardized Coefficients Std. Error	Standardized Coefficients Beta	t	Sig.	Collinearity Statistics Tolerance	Collinearity Statistics VIF
1	(Constant)	11.360	2.933		3.874	.000		
	No.employ	.563	.028	.846	20.380	.000	.761	1.314
	Price	-2.173	.564	-.160	-3.850	.000	.761	1.314

a. Dependent Variable: Revenue

10.12 Summary

In this chapter, we have described ordinary least squares (OLS) regression. It is a dependency technique where one or more independent variables (X_i) predict the value of one dependent variable (Y). It is one of the most widely used statistical methods that predict outcomes. The dependent variable must be measured at the interval or ratio level, though it is common to use Likert-type measures and treat them as approximately interval. That is, we treat them as a continuous variable. The independent variables can be measured at any level. Ordinal variables should be converted to nominal variables: one less nominal variable than the number of ordinal categories. There are seven classic assumptions for multiple regression analysis.

While important, OLS regression is fairly robust against violations of the assumptions.

References

Aguinis, H., Gottfredson, R. K., & Joo, H. (2013). Best-practice recommendations for defining, identifying, and handling outliers. *Organizational Research Methods, 16*(2), 270–301.

Hair, J. F., Black, W. C., Babin, B. J. & Anderson, R. E. (2014). *Multivariate Data Analysis, 7th edition, International edition,* Pearson Education Limited.

Cluster Analysis and Segmentation

11

Contents

11.1 Introduction

About 10 years ago, the International English School opened a campus in Uppsala. At the time, the Uppsala International School staff were very worried that they would lose a lot of their students when the new school opened. They were afraid that the two schools had the same offering and that students might change due to convenience or because the English School had new facilities. To investigate their fears, a professional marketing consultant along with one of the authors of this book asked the staff of the International School to describe what was unique with their educational offering. Their answer: "Uppsala International School does everything!"

This is a common mistake. For the sake of the example, think of the education as a product offering and the students as customers. What are the needs of students in the relevant age group in Uppsala? Can the needs be broken down into subgroups of students with unique needs? How does Uppsala International School's offering create value for the different subgroups? How does the offering differ with respect to the International English School?

Today, both schools flourish. The Uppsala International School staff learned that they have a unique and valuable offering that serves a unique and valuable subgroup of students who are not attracted by the International English School offering. You

J. E. Sallis et al., *Research Methods and Data Analysis for Business Decisions*, Classroom Companion: Business, https://doi.org/10.1007/978-3-030-84421-9_11

could say that each group of students forms a unique segment, or as we say in this chapter, a unique cluster.

In this chapter, we present different methods for segmenting data: in other words, methods to split data into smaller groups that, in statistical terms, we call *clusters*. In the two schools example, we could say that each school serves a different cluster of students. *A cluster is a natural grouping of homogenous cases in the data*, made in such a way that cases located in the same cluster are very similar to each other, while they are less similar to cases found in other clusters.

Intuitively, humans are good at classifying things. Already from early childhood, kids are able to distinguish between, for example, a swallow and duck. Despite the many characteristics that the two species have in common, such as feathers, a beak, and they can fly, they also have distinct differences. Cluster analysis acts in the same way by identifying important and defining properties that meaningfully distinguish groups. In contrast with traditional segmentation variables, cluster analysis finds natural groups already present, but hidden, in the data.

In marketing, clustering methods can be used for data exploration to better understand customers. For example, cluster analysis of questionnaire data will identify which customers may be grouped together or separated into segments. Respondents who have answered very alike on questions will naturally be considered close to each other, whereas respondents with dissimilar answers will be separated. The groups can then be identified by prominent and meaningful characteristics. Who are they? What are their preferences? How can they best be contacted?

Cluster analysis is often referred to as *the art of cluster analysis*. It is a highly subjective technique, requiring common sense and a strong logic for how the analysis is conducted. Results should be replicated and tested by, for example, using several clustering methods. Cluster analysis provides a glimpse into the world of data mining. With the advent of the Internet and digitalization, massive amounts of data provide the fodder for looking for patterns.

In the next section, we will consider a dataset with *k-variables and n cases*. Each case, for example, could be the responses from a respondent to a questionnaire. A case is represented as a vector with *k-dimensions* in a *k-dimensional* space. Think of the spatial distance between two cases as an expression of similarity (or dissimilarity) between them. Then, imagine finding *clouds*, which represent groups of cases with varying density within the data space. Finding clouds is a simple metaphor for how clustering methods work. We will discuss how to formalize the concept of equality as an expression of distance. We discuss two straightforward commonly applied clustering methods: agglomerative and divisive.

11.2 Similarities Between Groups in the Data

Clustering is about finding groups in data. An essential aspect is identifying which variables are most suitable for defining clearly discernable groups. This is not without its challenges. For example:

- Which bird is most similar to a duck, a crow or a penguin?
- According to the ability to fly, a crow is most similar.
- According to the ability to swim, a penguin is most similar.
- How, then, can we formalize the concept of similarity, or what we could call distance between two points?

A relatively simple and common way is called the *Euclidean distance*. In two dimensions, this is simply the distance between two cases on a two-dimensional plane. In layman terms, this is like the distance between two dots on a piece of paper. The distance between cases is the hypotenuse of a right-angled triangle, and is defined as the square root of the sum of the squared lengths of the two shorter sides. If we generalize this idea, the Euclidean distance between two cases i and j (in the k-dimensional space, representing k-variables) can be calculated by the formula:

$$Dist(i,j) = \sqrt{(x_{i1} - x_{ji})^2 + (x_{i2} - x_{j2})^2 + \ldots + (x_{ik} - x_{jk})^2} = \sqrt{\sum_k (x_i - x_j)^2}$$

This formula works best when applied to numeric, continuous variables. For example, for grouping companies, appropriate variables could be things such as turnover, tax rate, share price, and so on. An important point, however, is that the algorithms are sensitive to measurement magnitude. This means that variables with large values, like turnover, will have a greater influence on forming clusters than variables with small values, like return on investment. Even unit of measurement is important. For example, changing between currencies like NOK to USD will substantially influence the results. Potentially, these problems can be mitigated by standardizing all variables to a common unit and range so that all variables' magnitudes are relative to each other. However, some researchers would argue that this removes the natural weighting of the original variables, so it is not a foregone conclusion to always standardize the scales. As will become clear, often the best approach to clustering is to contrast the results from different approaches, like standardized or unstandardized variables, to evaluate the robustness of the solutions.

There are transformations for other types of variables, such as binary or nominal variables. Nominal variables with more than two categories can be treated in the same way as with regression (see Sect. 10.9 in the regression chapter for a more thorough discussion of dummy variables). Create *k-1* dummy variables for *k* categories. Binary variables with equally likely categories, like gender (50% men and 50% women), are a specific category. Binary variables with unequal categories may contain specific weightings that are important for interpretation. For example, 1% of infected people die from the COVID-19 virus, while 99% survive. This is an extremely different balance than the gender variable, though each can be equally important for forming clusters. In situations with such extreme imbalance, the researcher may choose, for example, to only cluster on survivors of the virus.

11.3 Two Branches of Cluster Methods

The aim of the clustering process is to ensure that the "clouds" of data we find are as *dense* as possible and at the same time *well-separated* from each other in the k-dimensional space, as defined by the variables. We distinguish between hierarchical and non-hierarchical clustering methods, where the hierarchical method builds a hierarchy of possible clusters solutions, and the non-hierarchical method clusters according to the number of groups specified by the user. A common strategy is to start with hierarchical clustering to determine the optimal number of clusters, and then move to K-means clustering to estimate centroids and cluster membership.

11.4 Hierarchical Clustering

Hierarchical clustering builds a *tree-like structure* from the data with the aim of finding the optimal number of clusters. The various methods can be roughly divided into two groups: agglomerative methods (see Fig. 11.1) and divisive methods (see Fig. 11.2). The agglomerative methods start with all cases separated, like the leaves on a tree, and then start joining similar cases successively until all the data is joined into one single cluster, which would be the stump of the tree. First, the two closest cases join. In the next iteration, the next closest cases join, then the next closest, and so on, and so on. Sometimes two single cases join, or a single case joins an already

Fig. 11.1 Agglomerative clustering

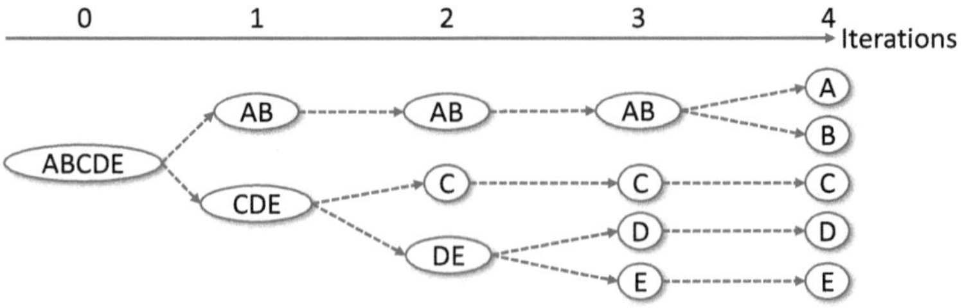

Fig. 11.2 Divisive clustering

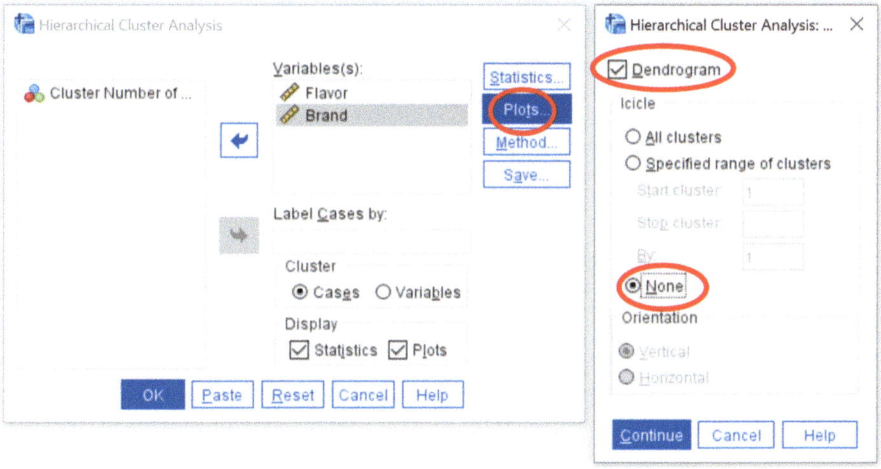

Fig. 11.3 Hierarchical cluster analysis

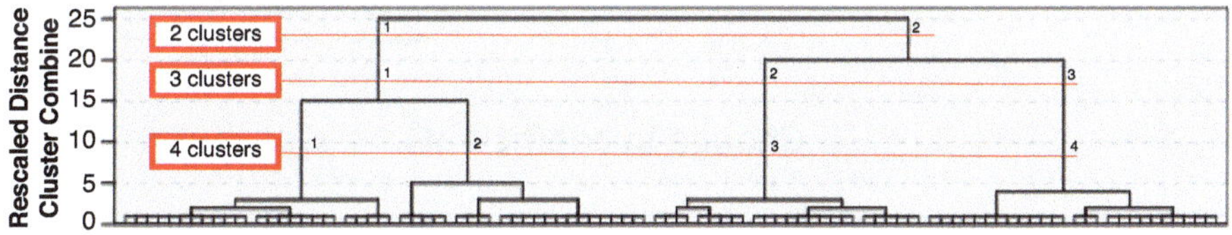

Fig. 11.4 Dendrogram

formed cluster. As the process continues, the goal is to stop when there is an optimal number of clusters to meaningfully represent the data.

Divisive methods have the same goal; however, they start with all observations together and then successively divide until all cases are separated.

The example in this chapter comes from a simulated dataset about beer consumption using two variables: flavor and brand loyalty. Imagine flavor as a measure from light-bodied beer (e.g., American) to heavy-bodied beer (e.g., Irish stout). Brand loyalty measures the degree to which drinkers are, or are not, loyal to brands. Also keep in mind that for the sake of simplicity, the data is ordinal with each variable measured on a 10-point scale. This violates the assumption of using continuous variables as input. However, it serves the purpose of demonstrating the method and can be useful in real contexts as well. For a discussion of treating ordinal variables as continuous, see Sect. 6.5.

Figure 11.3 shows the dialogue box for a hierarchical cluster analysis. In SPSS, open the Beer dataset. Choose *Analyze > Classify > Hierarchical Cluster. Put Flavour and Brand into the Variable(s) box. Choose Plots and select Dendogram, but deselect Icicle (choose none). Click continue, and OK.*

Figure 11.4 shows the *dendrogram* of the clustering process. To better fit the page, it is left-rotated from the original output. At the bottom where the process

Agglomeration Schedule

	Cluster Combined			Stage Cluster First Appears		
Stage	Cluster 1	Cluster 2	Coefficients	Cluster 1	Cluster 2	Next Stage
1	89	100	.000	0	0	11
2	79	99	.000	0	0	21
⋮						
94	11	16	6.600	89	83	97
95	1	2	7.731	80	90	98
96	8	10	9.822	92	86	97
97	8	11	35.748	96	94	99
98	1	4	48.563	95	93	99
99	1	8	61.122	98	97	0

61.1	
48.6	1.26
35.7	1.36
9.8	3.64
7.7	1.27
6.6	1.17

61.1/48.6 = 1.26

48.6/35.7 = 1.36

35.7/9.8 = 3.64

Change in coefficients

Fig. 11.5 Agglomeration schedule

starts, all cases are separate. Then, as the process continues, the analyst looks for cluster solutions that endure over several iterations. This is indicated by the length of the branches that are formed. In this example (moving from bottom to top), a 4-cluster solution is quite stable, then a 3-cluster solution, and finally a two-cluster solution. The conclusion from the dendrogram would then be to run the 4-, 3-, and 2-cluster solutions in K-means clustering, and then determine which one has the best substantive meaning.

Another diagnostic tool is to look at the agglomeration schedule and calculate the change in the coefficients (see Fig. 11.5). Start by dividing the bottom coefficient with the one above it (2-cluster solution), and then the next coefficient with the one above it (3-cluster solution), and carry on looking for where substantial changes occur. In this example, 3.64 is substantially bigger than any other changes around it, suggesting a 4-cluster solution for testing in a K-means cluster analysis.

So far, the evidence is suggesting a 4-cluster solution. Given the subjective character of cluster analysis, at this point it could be wise to, for example, *standardize* the variables to see whether the solution is robust. Alternatively, test *different clustering methods* or *distance measures* between cases. With the hierarchical cluster dialogue box open, *click on Method, and make your choices (see* Fig. 11.6*).*

Once satisfied with the number of clusters to extract, move on to K-means clustering.

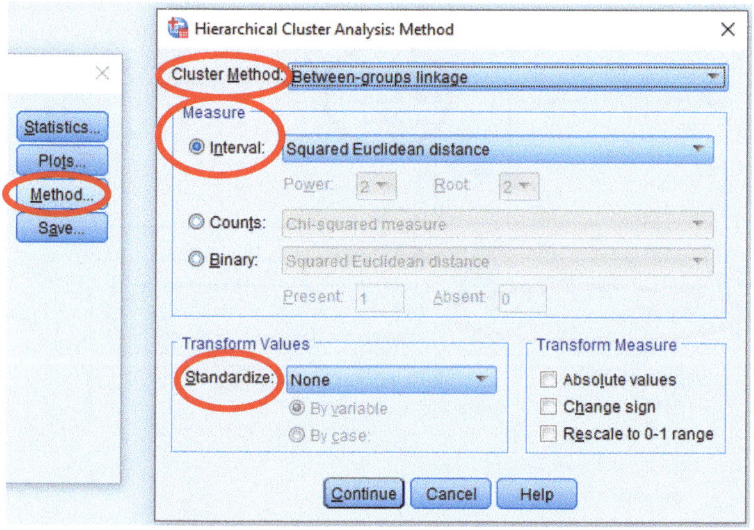

Fig. 11.6 Cluster methods, distance measures, and standardization

11.5 Non-hierarchical Clustering (K-Means)

K-Means clustering is a classic example of a non-hierarchical clustering method. The algorithm uses the parameter *K*, which represents the number of clusters to be derived, so it must not be confused with *k*, which often refers to the number of variables in the dataset. The input is a dataset with a number of (numeric) variables, the result being a set of clusters grouped around *centroids*, which are the center cases in the clusters. The chosen distance measure determines the degree of similarity *as distance between cases* in the k-dimensional space. The procedure is as follows: Continuing with the Beer dataset in SPSS, choose *Analyze > Classify > K-Means Cluster. Put Flavour and Brand into the Variable(s) box. Input the number of clusters to be extracted. Click OK (see* Fig. 11.7*).*

Note that *Cluster Centers (centroids)* can be read from a file. Advanced users may find it advantageous to specify their own centroids.

The next step is to *choose Save, and tick Cluster membership, continue, and OK (see* Fig. 11.8*).*

The software runs an iterative process until the algorithm converges on a stable solution. Table 11.1 shows the number of cases in each cluster. This example is quite balanced with 25, 27, 23, and 25 cases in each cluster, respectively.

Table 11.2 shows the cluster centers identified by the software. Imagine a two-dimensional plane with Flavor on the Y-axis and Brand on the X-axis, and 10-point scales on both axes. This is facilitated by graphing the solution and identifying cluster membership.

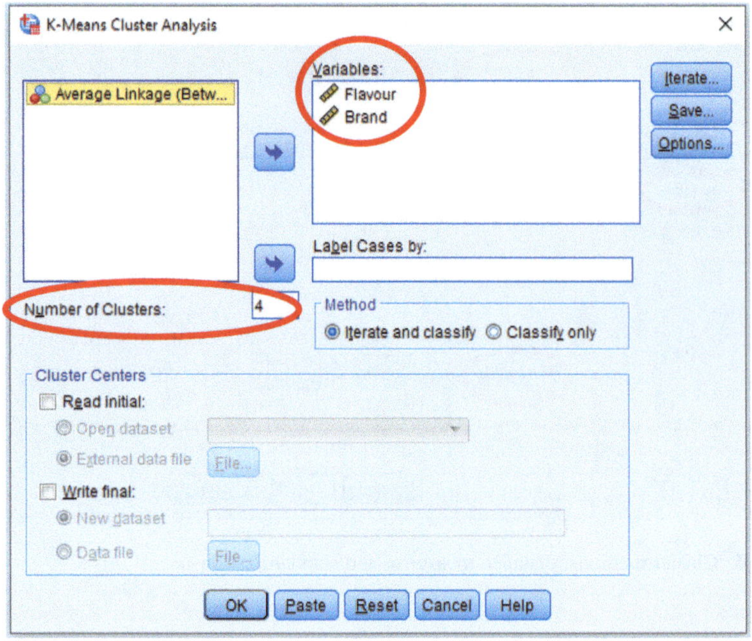

Fig. 11.7 K-means cluster analysis

Fig. 11.8 Cluster membership

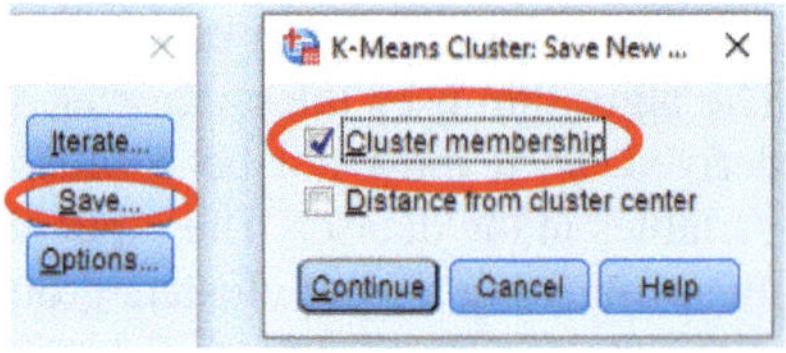

Table 11.1 Cases in each cluster

Number of cases in each cluster		
Cluster	1	25.000
	2	27.000
	3	23.000
	4	25.000
Valid		100.000
Missing		0.000

Table 11.2 Cluster centroids

Final cluster centers				
	Cluster			
	1	2	3	4
Flavor	3	9	9	2
Brand	9	3	9	2

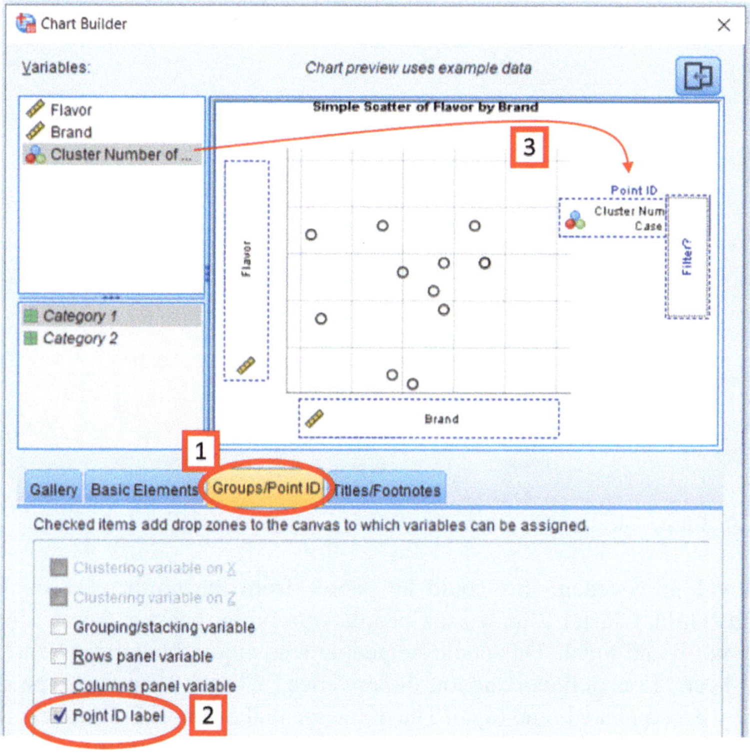

Fig. 11.9 Chart builder for cluster membership

To create a graph in SPSS, *choose Graphs > Chart builder > Scatter, then drag and drop the Simple Scatter into the graph window. Drag and drop Flavor to the Y-axis and Brand to the X-axis. Choose Groups/Point ID, and tick Point ID Label. Drag and drop the Cluster Number variable to the Point Label Variable. Click OK (see* Fig. 11.9*).*

Figure 11.10 shows the graph with a 4-cluster solution. The cluster centroids from Table 11.2 are also indicated. This is an excellent solution with all four clusters distinctly separated and grouped in different quadrants. Keep in mind that the data is simulated to demonstrate a good solution.

11.6 Interpretation and Further Use of the Clusters

In the interpretation process, variables that were included in the clustering or other relevant variables from the same dataset may help with interpretation. The different clusters should be evaluated and documented so that the results take on a pragmatic, substantive meaning. In the beer example there are four different clusters that need to be interpreted. Cluster 1 represents people who prefer light flavored beer and are

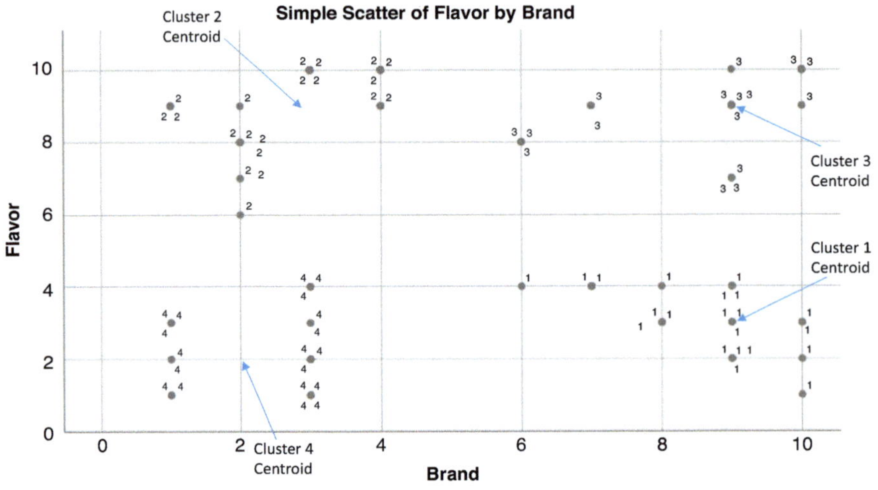

Fig. 11.10 Cluster membership graph

brand loyal. In Sweden, this could be people from the north who are loyal to Norrlands Guld. Cluster 2 represents people who prefer heavier flavored beer, but who are not brand loyal. This could be people who enjoy the current trend in craft brewery beer: "Lots of flavor and lots of breweries." Cluster 3 represents people who like heavy beer and are brand loyal. The quintessential example is Guinness Stout, to whom the Irish are very loyal. Cluster 4 represents people who like light bear and who are not brand loyal. These are people who may like American or Mexican beer, but otherwise do not care.

The cluster descriptions demonstrate how important it can be to have other variables, for example, demographic variables (age, gender, etc.), to build specific profiles around each segment (cluster). If nationality was measured, or home address, we could test the suppositions about Irish people and Guinness, and Swedish Northerners and Norrlands Guld.

Essentially, clustering is a rather subjective method of exploring data. Clusters provide the opportunity to divide the data and apply other statistical methods, for example, testing for group differences in regression models. The assumption is that the clusters represent important groups that deserve representation in the data analysis. It may be advantageous to model the groups separately. For example, create "Behavioral Clusters" that can then be cross referenced with "Demographic Clusters," which may provide valuable insights, or even provide valuable input for further analysis.

11.7 Summary

This chapter has covered the art of cluster analysis, which is a way for segmenting cases in data. Two specific methods were covered: agglomerative clustering and K-means cluster. Agglomerative clustering is primarily used to determine the optimal number of clusters, whereas K-means clustering is used to estimate cluster centers (centroids) and cluster membership for the cases in the data.

A cluster is a natural grouping of data, where the groups are as homogeneous as possible, while at the same time being as distinct as possible from each other. In marketing, clustering methods are used as a method of data exploration to better understand customers. The chapter example was on a simulated dataset for beer consumption. The results identified four distinct cluster, or what would be called segments in marketing.

Results of cluster analysis can provide meaningful input for further statistical analysis using other techniques.

Factor Analysis 12

Contents

12.1 Introduction

Would you let someone measure your intelligence with one question? What if you are good at algebra, but bad at geometry? Perhaps you are gifted with language and poor at mathematics? How can a single question capture all the dimensions of human intelligence to arrive at a reasonable conclusion? In research method terms, can human intelligence be validly measured with a single question? Our answer is no.

Exploratory factor analysis was originally designed to explore "general intelligence," as indicated by student performance in different courses. Spearman (1904) believed that observed performance in different courses was correlated and that the correlations could be used to reflect a latent construct of general intelligence.

Many constructs are multifaceted, meaning they have more than one dimension. If we were doing research on human intelligence, we might start by looking at various theoretical definitions, and from there consider how to operationalize it. Most likely, any definition would acknowledge that human intelligence is multifaceted. Validly measuring it necessitates measuring the underlying dimensions that together indicate intelligence. In Chap. 6, we defined a *latent construct* as a phenomenon that is not directly observable. Though we may jump to conclusions about another person's intelligence (e.g., when driving a car), in fact, intelligence is a

J. E. Sallis et al., *Research Methods and Data Analysis for Business Decisions*, Classroom Companion: Business, https://doi.org/10.1007/978-3-030-84421-9_12

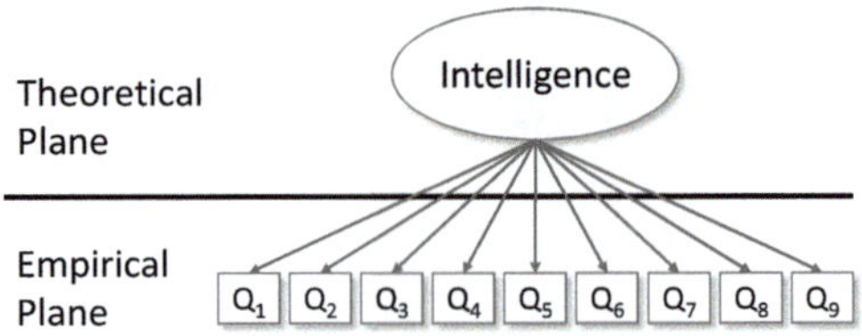

Fig. 12.1 Latent construct of intelligence with nine indicators

Table 12.1 Grouping variables versus grouping subjects

Factor Analysis: 9 Indicators (variables) → 1 Factor (new variable)

Cluster Analysis: Subjects

	V_1	V_2	V_3	V_4	V_5	V_6	V_7	V_8	V_9	Intelligence
S_1	7	5	3	4	3	6	6	5	5	4.89
S_2	4	5	2	3	4	6	5	4	5	4.22
S_3	6	4	1	3	4	7	5	3	6	4.33
S_4	4	5	2	5	4	5	6	3	6	4.44
S_5	7	5	2	4	3	6	6	4	4	4.56
S_6	5	2	4	3	2	6	5	5	5	4.11
S_7	6	6	3	4	3	5	7	5	6	5.00
S_8	5	4	1	5	3	7	5	6	5	4.56

multifaceted latent construct. Figure 12.1 shows a latent construct for intelligence measured by nine questions.

In Chap. 6, we discussed construct validity, which is made up of convergent validity and discriminant validity. *Construct validity* is the degree to which measures are related to a specific construct and not related to other constructs. *Convergent validity* is the extent to which specific measures converge on a construct, and *discriminant validity* is the extent to which the same measures do not converge on other constructs. In the current example, the construct is intelligence. Factor analysis can be used to measure the convergence of measures on a construct, and whether the measures discriminate from other constructs. Because factor analysis is often used to assess measurement validity, the individual variables are often called *indicators*, because they indicate a multidimensional latent construct.

Table 12.1 shows how cluster analysis, which we discussed in Chap. 11, groups subjects. These could be respondents to a survey. In contrast, *factor analysis* determines which indicators (variables) may be grouped together and how well they discriminate from each other. There is a method called *Q-type factor analysis* that, like cluster analysis, groups respondents. However, we do not discuss it in this text and refer the reader to cluster analysis for grouping subjects. Though it is rarely referred to in this way, exploratory factor analysis for grouping variables is called *R-type factor analysis*.

Factor analysis is generally known as a *data reduction technique*. It takes a set of variables and determines whether and how they can be reduced to a fewer number of factors. Our intelligence example suggests that nine questions, which we call indicators, represent intelligence. In other words, the nine variables form a single factor. If we establish that the nine indicators validly and reliably indicate the single factor of intelligence, we have two choices:

1. Keep the indicators separate and move on to other analysis techniques, like structural equation modeling, where it is possible to model latent constructs and their indicators. We discuss this near the end of this chapter under *confirmatory factor analysis*.
2. Add the indicators together to form a new single "aggregated" variable. In Table 12.1, the new "intelligence" variable (factor) to the right is an average of the nine indicators of intelligence. The new variable could be used in subsequent analyses, like for example regression.

In the remainder of the chapter, we will explain the difference between exploratory factor analysis, principal component analysis, and confirmatory factor analysis. We will demonstrate how to run exploratory factor analysis and reliability analysis. The general purpose is to show how to reduce several variables into factors (new variables). Specifically, we frame the chapter around establishing construct validity and reliability for several measures in a questionnaire.

Exploratory factor analysis with maximum likelihood estimation is the method we will explain. Strictly speaking, the data is supposed to be at the interval or ratio level. However, given its extensive use to assess questionnaire data with Likert scales, we will use questionnaire data based on 7-point scales. As discussed earlier, we will assume this data to be *approximately interval*.

12.2 Exploratory Factor Analysis

There are several different methods to *extract factors* from the data. The extraction methods can be divided into two categories: *principal components analysis* and all other methods (including maximum likelihood) that are grouped under the name *exploratory factor analysis*. They are distinguished by the way they treat the variance in indicators. Figure 12.2 shows three kinds of variance that account for the total variance when assessing indicators of latent constructs.

For the nine indicators of the intelligence construct, each indicator will have *common variance* that it shares with the other indicators. *Specific variance* is unique to only one indicator, and *error* is unexplained variance. *Factor loadings* are the correlation between the indicator and the latent construct. They show the degree of common variance an indicator shares with all other indicators in the analysis. When

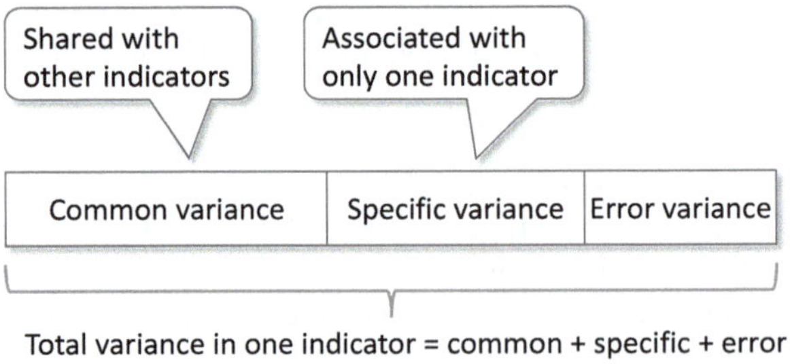

Fig. 12.2 Variance between variables

Table 12.2 Factor (component) thresholds

Factor (component) thresholds	Absolute value of factor loading
Poor	0.32 – 0.44
Fair	0.45 – 0.54
Good	0.55 – 0.62
Very good	0.63 – 0.70
Excellent	0.71 – 1.00

a factor loading is "high," it shares a lot of common variance with that factor. A good result would be for indicators of a latent construct to have high loadings on the same factor and low loadings on other factors.

The thresholds for the magnitude of factor loadings are debatable. Exploratory factor analysis is somewhat subjective, so it is important to build a chain of evidence for the decisions when assessing measures. The thresholds in Table 12.2 are from Tabachnick and Fidell (2013) and Comrey and Lee (1992). Note that the thresholds are also applicable to principal component analysis.

Exploratory factor analysis *only uses common variance* for extracting factors. This is based on the assumption that the common variance represents a significant proportion of the total variance and that the variation in the observed indicators is due to the presence of one or more latent variables (factors) that have a *causal influence* on the observed indicators. The causal influence is why the indicators are referred to as reflective measures. *Reflective measures* are like the symptoms of an underlying latent construct. *Causal measures* are the underlying reason (or cause) of a latent construct. As a metaphor, consider the COVID-19 virus. The symptoms, like fever, body aches, shortness of break, and loss of smell and taste, are reflective indicators of a person having the virus. A doctor may diagnose the virus based on a sample of symptoms. The virus itself is a causal indicator of the virus. Without this cause being present, it is impossible for the patient to contract COVID-19.

As a rule of thumb, where possible we recommend using reflective indicators. The correct causal indicators *must* be included in the measure; otherwise, it is not valid. It is not possible to cause COVID-19 without the virus. Causal indicators are not necessarily correlated, so exploratory factor analysis and reliability analysis are not optimal for establishing construct validity and unidimensionality. Valid, but uncorrelated, causal indicators may be incorrectly rejected. For reflective measures, it suffices to have a sample of indicators to represent the dimensions of the latent construct. There are many methods for assessing the construct validity of reflective measures. In sum, reflective measures are easier to work with. For a detailed discussion, see Bollen and Lennox (1991).

12.3 Principal Component Analysis

Principal component analysis, like exploratory factory analysis, is used to analyze the underlying structure of the data to see whether and how the indicators/variables can be reduced to a smaller number. It differs in that it uses both common and specific variance to extract a solution and the related indicators are called components. It seeks to find the linear combination between indicators that extracts the most variance in the data. It does not assume that the latent construct causes the indicators, so it does not assume that the indicators are correlated (Table 12.3).

12.4 Running Exploratory Factor Analysis

As an example of exploratory factor analysis, we turn to the *Hotel* data. Assume that management took the initiative to make a questionnaire with 20 questions measuring four latent constructs. Each construct had five indicators, measured on 7-point scales from completely disagree (1) to completely agree (7). The constructs were:

Table 12.3 Exploratory factor analysis versus principal component analysis

Exploratory factor analysis	Principal component analysis
• Assumes causal influence on the observed indicators/variables	• Assumes a linear combination between observed indicators/variables
• Assumes substantial correlation between observed variables	• Does not assume correlation between observed variables
• Uses common variance to extract a solution	• Uses common and specific variance to extract a solution
• The latent constructs are called factors	• The latent constructs are called components

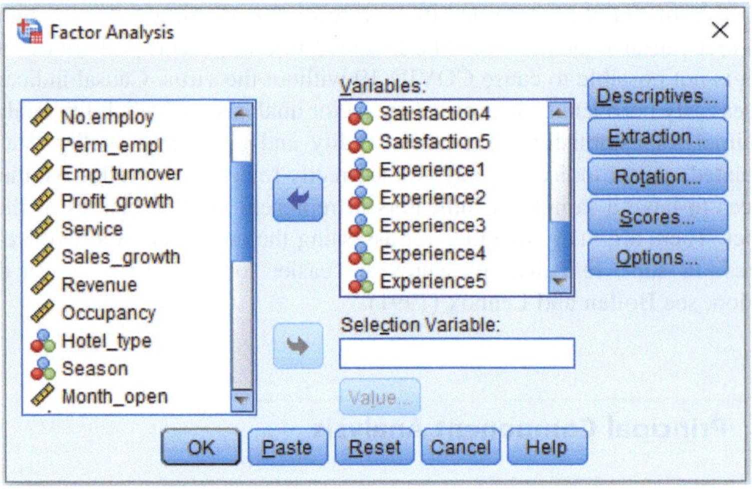

Fig. 12.3 Exploratory factor analysis dialogue box

Satisfaction, with questions measuring dimensions of satisfaction.

Offering diversity, with questions about facilities like pools, spa, training room, and restaurants

Customer experience, with questions covering staff friendliness, cleanliness, updated facilities, and room decor

Location, with questions on things like airports, public transportation, convention facilities, and tourist sites.

Determining the Factorability of the Data

The first step in running the analysis is to determine the factorability of the data. Since factor analysis uses correlations to converge on a solution, the input variables, in this case indicators from the questionnaire data, must have a substantial number of fairly large correlations. Though we do not show the correlation matrix between indicators here, we can confirm that a little more than half of the possible correlations are significant, with the largest correlation coefficient at 0.817.

The next step is to input the data into the factor analysis software and request the appropriate statistics. In SPSS, open the *Hotel* data and . Choose *Analyze > Dimension Reduction > Factor. Put the 20 indicators for the four latent variables (Offering, Location, Satisfaction, and Experience) into the Variable(s) box (see Fig. 12.3).*

Choose Descriptives > tick KMO and Bartlett's test of sphericity (see Fig. 12.4).

KMO, which is the *Kaiser–Meyer–Olkin* test, is a measure of suitability of the data. The value will be from 0 to 1, with 1 being perfect. Proving that statisticians do have a sense of humor, Table 12.4 shows the threshold values and associated descriptions.

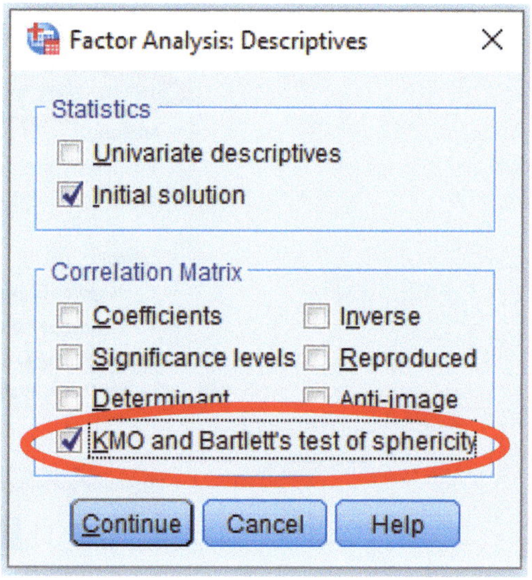

Fig. 12.4 Descriptive statistics and KMO

Table 12.4 KMO interpretation

Description	KMO thresholds
Unacceptable	0.00 – 0.49
Miserable	0.50 – 0.59
Mediocre	0.60 – 0.69
Middling	0.70 – 0.79
Meritorious	0.80 – 0.89
Marvelous	0.90 – 1.00

Extraction Method and Choosing the Number of Factors to Extract

Choose Extraction > and using the dropdown box at the top, choose Maximum liklihoood as the Method (see Fig. 12.5).

It is beyond the scope of this book to explain other extraction methods, except for the brief description of principal component analysis.

In Fig. 12.5, in the bottom half of the dialogue box there is a section called, "Extract." The default method, which is the most common when exploring data, is to extract the number of factors when the *Eigen value is above 1*. The other method is to choose a *fixed number of factors to extract*. Although this is called exploratory factor analysis, often the researcher has an idea of how many factors to expect. In the current example, we expect 4 factors representing the latent variables being measured. You should set the number to 4 and move on with the analysis.

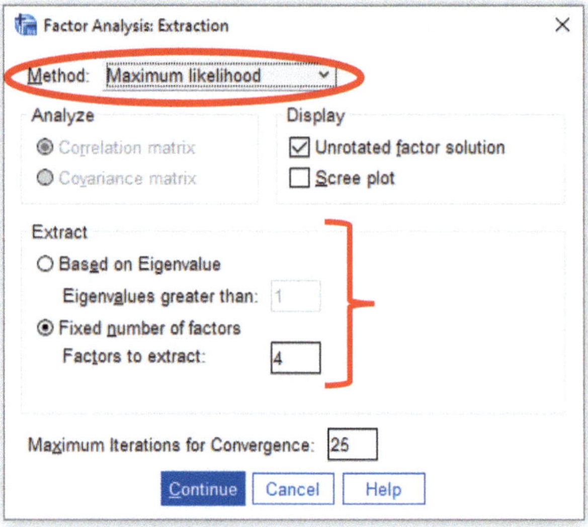

Fig. 12.5 Extraction method and number of factors

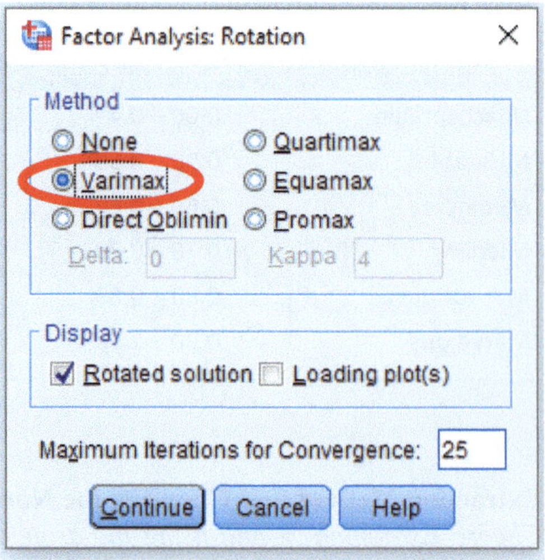

Fig. 12.6 Factor rotation

Eigen values express the explained variance of each additional factor in a factor analysis. The maximum possible number of factors equals the maximum number of variables. In our example, this means 20 variables equals 20 possible factors. Given the purpose of reducing the variables into a smaller number of factors, the question becomes, how many factors are optimal? Using the Eigen value criterion, the researcher extracts the number of factors indicated by the Eigen value being above 1.

Factor Rotation

To facilitate the interpretation of the results, the research must decide on some form of factor rotation method. *Choose Rotation > and in the Factor analysis:Rotation dialogue box, tick Variamax, and continue (see Fig. 12.6).*

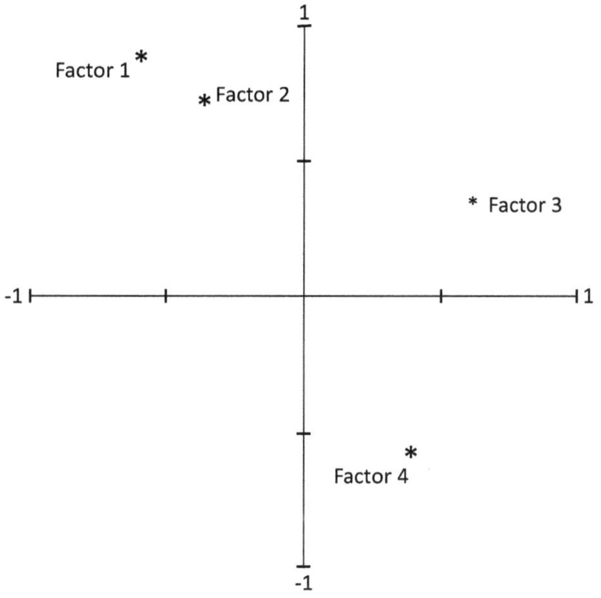

Fig. 12.7 Unrotated factor solution

There are two basic types of factor rotation: orthogonal and oblique. *Orthogonal rotation* means to rotate two axes in a two-dimensional space, keeping the origins at a 90-degree angle. It assumes that the underlying factors are not correlated. *Oblique rotation* means to rotate two axes in a two-dimensional space; however, the origins are allowed to change their angle from 90 degrees. It assumes that the factors are correlated. The purpose of either rotation is to make the results easier to interpret.

Figure 12.7 shows four factor centers plotted in an unrotated two-dimensional space. The axes do not pass closely through the factors, making the factor loadings difficult to interpret. To understand what we mean, refer to Table 12.8, which shows the unrotated factor loadings.

The blue dotted lines in Fig. 12.8 show an orthogonal rotation (keeping the axes at 90 degrees), such that the axes pass as close as possible through the factors. This makes interpreting the factor loadings much easier. Refer to Tables 12.9–12.11, which show a much clearer picture of how the indicators load on the factors.

Suppress Small Values

Exploratory factor analysis estimates factor loadings for all variables on all factors, including all small factor loadings. This makes reading the tables somewhat cumbersome. Table 12.2 shows factor loading thresholds for interpretation. According to the table, the lowest factor loading to be interpreted would be 0.32 (poor). For the sake of readability, we suggest: *Choose Options > tick Suppress small coefficients > and set Absolute value below 0.32, then click continue (see Fig. 12.9).* When running the factor analysis, you may find it practical to adjust the value according to how you want to communicate the findings.

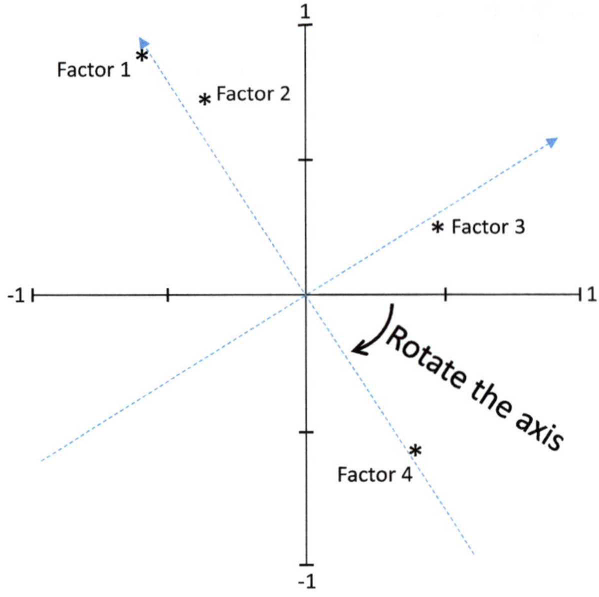

Fig. 12.8 Rotated factor solution

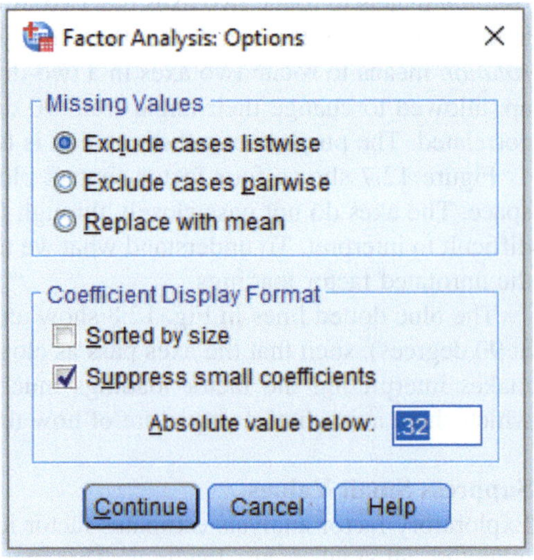

Fig. 12.9 Suppress small coefficients

Exploratory Factor Analysis Output

Table 12.5 shows the results of the KMO test. With a value of 0.835, the data is meritorious for factorability. If the KMO was considerably lower, we would look for problematic indicators to remove.

Table 12.5 KMO test

KMO and Bartlett's Test

Kaiser-Meyer-Olkin Measure of Sampling Adequacy.		.835
Bartlett's Test of Sphericity	Approx. Chi-Square	1094.026
	df	190
	Sig.	.000

Communalities, shown in Table 12.6, indicate how correlated individual indicators are with the rest of the indicators. An indicator with a low communality may not load well on any factor. There is no broadly accepted value indicating low. However, our rule of thumb is to be concerned when a communality is below 0.5. Remember, exploratory factor analysis is a somewhat subjective technique. The decision whether to eliminate an indicator should be made based on overall evidence rather than a hard rule of, for example, a communality below 0.5.

In this example, when estimating the communalities the software gave the warning, "One or more communalitiy estimates greater than 1 were encountered during iterations. The resulting solution should be interpreted with caution." In factor analysis it is fairly common to have convergence issues for various reasons. This is why, "getting to know your data," is so important. Often, problems are solved by removing problematic cases or variables. When you encounter problems, if you know your data well, then you will probably have ideas around how to proceed. In this example, the solution converged so we have enough information to proceed. Nevertheless, interpret the commonalities with caution. The commonalities that are 0.999 (see the blue arrows) may indicate a problem with the two indicators.

Also in Table 12.6, note that the first three indicators (Offering1, Offering2, and Offering3) have low communalities. As we continue with the analysis, keep in mind that these indicators may be problematic. From a pragmatic perspective, the researcher could look at the questions measuring this construct to try and understand whether there may be a problem.

Table 12.7 shows part of the *Total Variance Explained* table. In our example, there are four latent constructs, so we are seeking a 4-factor solution. It is worth noting that the total Eigen value suggests a 5-factor solution. This is not a serious problem. The total Eigen value for the 5th factor is 1.083, which is close to the cutoff of 1.00. Again, this is just more evidence to take into consideration during the analysis process.

It is also worth noting from Table 12.7 that the total explained variance for a 4-factor solution is 68.658%.

Table 12.8 shows the unrotated factor solution with suppressed values below 0.32. The factor loadings appear randomly distributed about the table, with the majority on factor 2. It is impossible to interpret the proper factor structure from this table.

Table 12.6 Communalities

Communalities[a]

	Initial	Extraction
Offering1	.368	.218
Offering2	.419	.279
Offering3	.390	.260
Offering4	.603	.999
Offering5	.556	.453
Location1	.594	.592
Location2	.733	.723
Location3	.730	.797
Location4	.754	.665
Location5	.717	.690
Satisfaction1	.629	.587
Satisfaction2	.779	.782
Satisfaction3	.788	.797
Satisfaction4	.839	.846
Satisfaction5	.748	.703
Experience1	.560	.501
Experience2	.672	.706
Experience3	.675	.687
Experience4	.607	.555
Experience5	.519	.999

Extraction Method: Maximum Likelihood.

Table 12.9 shows the rotated factor solution for all 20 indicators. Each factor represents a latent construct, and we are looking for patterns of which indicators load on which factor. The way the factor loadings are spread suggests that Factor 1 is Satisfaction, Factor 2 is Location, Factor 3 is Offering, and Factor 4 is Experience. The loadings in red circles are associated with the wrong factor. This means, for example, that question 1 of the Offering scale fits better with the Loacation scale. By examining the questionnaire, the researcher might understand why.

Table 12.7 Total variance explained

Total Variance Explained

Factor	Initial Eigenvalues Total	% of Variance	Cumulative %
1	7.604	38.021	38.021
2	2.564	12.821	50.842
3	1.974	9.871	60.714
4	1.189	5.945	66.658
5	1.083	5.413	72.071
6	.847	4.234	76.305
7	.798	3.990	80.295
8	.592	2.959	83.254
9	.519	2.597	85.851
10	.419	2.097	87.948
11	.416	2.082	90.030
12	.354	1.770	91.800
13	.345	1.726	93.526
14	.286	1.429	94.955
15	.241	1.203	96.158
16	.216	1.081	97.239
17	.178	.891	98.130
18	.161	.806	98.936
19	.115	.573	99.509
20	.098	.491	100.000

Extraction Method: Maximum Likelihood.

Offering 3 does not have any factor loadings above 0.32, meaning it should probably be dropped from the analysis. The factor loadings in green circles are called cross-loadings. This means that they load on two factors. It also means that they do not have discriminant validity. That is, they do not discriminate between which factor they load on. All the values are larger on Factor 4, though they also want to load on Factor 1.

Continuing with the analysis, the researcher may want to try several different models. Often, it is a good idea to check latent variables in pairs. Table 12.10 shows the final solution. Note that there are still three cross-loadings circled in green.

Table 12.8 Unrotated factor matrix

Factor Matrix[a]

	Factor			
	1	2	3	4
Offering1			.356	
Offering2	.396			
Offering3		.436		
Offering4	.999			
Offering5	.631			
Location1		.612	.410	
Location2		.645	.510	
Location3		.486	.717	
Location4		.648	.480	
Location5		.520	.632	
Satisfaction1		.735		
Satisfaction2		.833		
Satisfaction3		.773	-.323	
Satisfaction4		.846		
Satisfaction5		.762	-.331	
Experience1		.681		
Experience2		.674		.391
Experience3		.692		.462
Experience4		.642		.393
Experience5		.443		

Extraction Method: Maximum Likelihood.

a. 4 factors extracted. 6 iterations required.

Satisfaction 5 clearly has the highest loading on Factor 1 (0.733), Experience 2 loads much higher on Factor 3 (0.667), and Experience 4 clearly loads higher on Factor 3 (0.625).

By increasing the threshold to suppress small coefficients to 0.43, the final output of the exploratory factor analysis is straightforward to interpret. There are clearly

Table 12.9 Rotated factor matrix

Rotated Factor Matrix[a]

	Factor			
	1	2	3	4
Offering1		.438		
Offering2			.422	
Offering3				
Offering4			.993	
Offering5			.621	
Location1		.666		
Location2		.784		
Location3		.884		
Location4		.740		
Location5		.778		
Satisfaction1	.701			
Satisfaction2	.846			
Satisfaction3	.855			
Satisfaction4	.880			
Satisfaction5	.777			
Experience1	.625			
Experience2	.499			.591
Experience3	.419			.664
Experience4	.435			.579
Experience5	.407			

Extraction Method: Maximum Likelihood.
Rotation Method: Varimax with Kaiser Normalization.

a. Rotation converged in 6 iterations.

four factors that have indicators (variables) with high factor loadings and there are no serious cross-loadings. The lowest factor loading is Experience 4 (0.625), which according to the thresholds in Table 12.2 is good. All other factor loadings are in the very good to excellent range. This means that the indicators *converge* on factors and *discriminate* from other factors. Together, this is good evidence of *construct validity*.

In case you noticed that during the modeling process, the factor loadings moved around with respect to which factor they loaded on. In Table 12.9, Offering was

Table 12.10 Rotated factor matrix with cross-loadings

Rotated Factor Matrix[a]

	Factor			
	1	2	3	4
Offering4				.997
Offering5				.640
Location1		.649		
Location2		.770		
Location3		.888		
Location4		.749		
Location5		.779		
Satisfaction1	.659			
Satisfaction2	.830			
Satisfaction3	.838			
Satisfaction4	.872			
Satisfaction5	.733		.365	
Experience2	.426		.667	
Experience3			.701	
Experience4	.346		.625	

loading on factor 3, while Experience was loading on Factor 4, and in subsequent tables they switched places. This is not a problem, it is a function of how the factors are derived. It has no substantive meaning for the analysis, so just ignore it (Table 12.11).

To check the robustness of the results, it can be a good idea to run the same analysis using different estimation methods. Though not shown, first, we replicated the analysis using principal components as the extraction method. Of course, the numbers differed slightly; however, the conclusions were the same. Then, using Maximum Likelihood to extract the factors, we changed the rotation method to Direct Oblimin. Again, the conclusions were the same. Of all the possible factoring methods available in SPSS, the only time we reached a different conclusion was when we chose Quartimax as the rotation method. There were issues with cross-loadings. We conclude that our findings are robust.

In factor analysis, there are generally accepted conventions for how to graphically represent factor models. Figure 12.10 shows the basic conventions.

Table 12.11 Final rotated factor matrix

Rotated Factor Matrix[a]

	Factor			
	1	2	3	4
Offering4				.997
Offering5				.640
Location1		.649		
Location2		.770		
Location3		.888		
Location4		.749		
Location5		.779		
Satisfaction1	.659			
Satisfaction2	.830			
Satisfaction3	.838			
Satisfaction4	.872			
Satisfaction5	.733			
Experience2			.667	
Experience3			.701	
Experience4			.625	

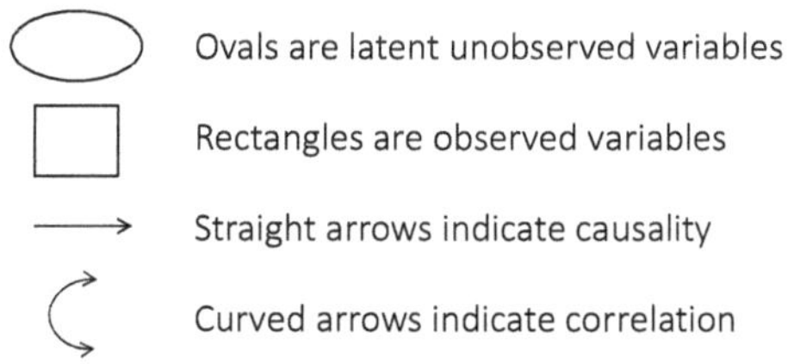

Fig. 12.10 Drawing conventions

Figure 12.11 shows how the rotated factor matrix can be shown as an exploratory factor analysis model. Error, by convention, is represented by the Greek letter delta (δ). The arrows show the causal reflective relationship between all the indicators and all the latent constructs. The solid black arrows show the "large" factor loadings, and the blue dotted arrows show the small factor loadings.

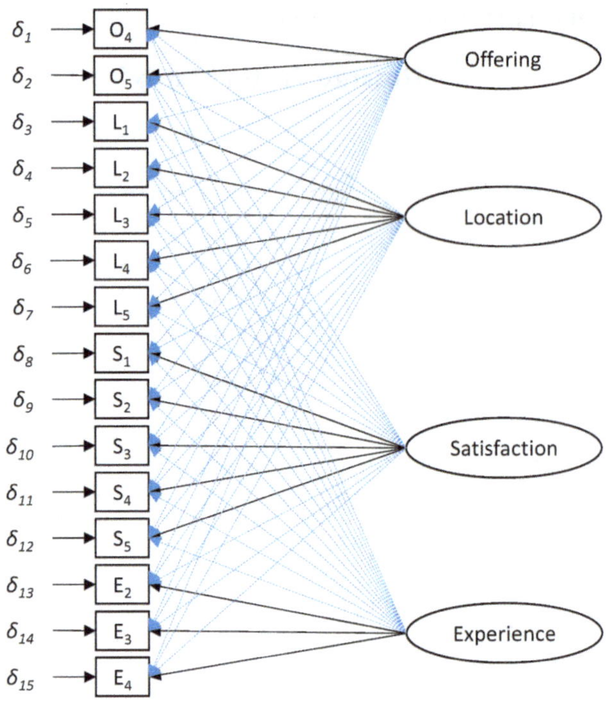

Fig. 12.11 Exploratory factor analysis model

12.5 Unidimensionality

In Chap. 3, we said *Validity* refers to how well we measure what we intend to measure, and *Reliability* refers to the consistency of how well we measure something. Both, from a measurement perspective, establish the *unidimensionality* of the measures of a latent construct. When assessing measures, it can be a good idea to go back and forth between testing construct validity and reliability until deciding on the best possible measures of a latent construct.

An important rule of thumb is to avoid being blinded by and just chasing numbers. Unless the analysis is truly exploratory, without any ex ante idea of the relationship between variables, theory should always play a role in decisions. The best, *by the numbers measures*, may not be the best measures of a construct. If a latent construct has clearly defined subdimensions, then they must be captured in the measures. If the factor analysis or reliability analysis shows superior results without all the dimensions, though statistically valid and reliable, the construct lacks *content validity*. It does not properly measure the theoretically defined construct.

In Chap. 8, we discussed Cronbach's alpha and went through an example. Nevertheless, given its relationship to scale development using exploratory factor analysis, we briefly show it again here. We estimated Cronbach's alpha for the final measures of the latent constructs. The results are in Table 12.12. According to

Table 12.12 Cronbach's alpha results

Latent construct	Cronbach's alpha
Offering (4 & 5)	0.782
Location (1-5)	0.902
Satisfaction (1-5)	0.926
Experience (2-4)	0.825

Nunnally's (1978) cutoff criteria of 0.7, the measures are reliable. Together with the exploratory factor analysis, we can conclude that the measures are unidimensional.

12.6 Confirmatory Factor Analysis

When we explained exploratory factor analysis, we used an example where we, ex ante, had an idea of what factor structure to expect. This is typical when testing established measures for latent constructs. In plain terms, we knew which questions would measure which constructs. Despite that we used exploratory factor analysis to confirm our beliefs, it is not confirmatory factor analysis. Confirmatory factor analysis uses the chi-square distribution to test the hypothesis of whether the imposed factor structure fits the data. Karl Jöreskog started developing the method at Uppsala University during the 1960s. It is part of a larger group of methods in structural equation modeling (SEM). Jöreskog and Dag Sörbom developed SEM software, called LISREL, which remains popular today. For a detailed description of structural equation modeling in LISREL, we refer to Jöreskog, Olsson, and Wallentin (2016).

The Measurement Model

The first difference to note when comparing it with exploratory factor analysis is that the researcher imposes a structure on the data, rather than allowing the program to suggest a solution. In Fig. 12.12, the same model is proposed that is shown in Fig. 12.11. The obvious difference is the lack of factor loadings from all latent constructs to all indicators. This explains the disappearance of the dotted blue lines. The other difference is that the latent variables are allowed to correlate. This allows the researcher to assess discriminant validity by testing the correlations between latent constructs.

The Structural Model

The structural model in confirmatory factor analysis allows for the specification of causal relationships between latent constructs. Anderson and Gerbing (1988) established the practice of first estimating the measurement model, and then estimating the structural model. Figure 12.13 shows the structural model we could

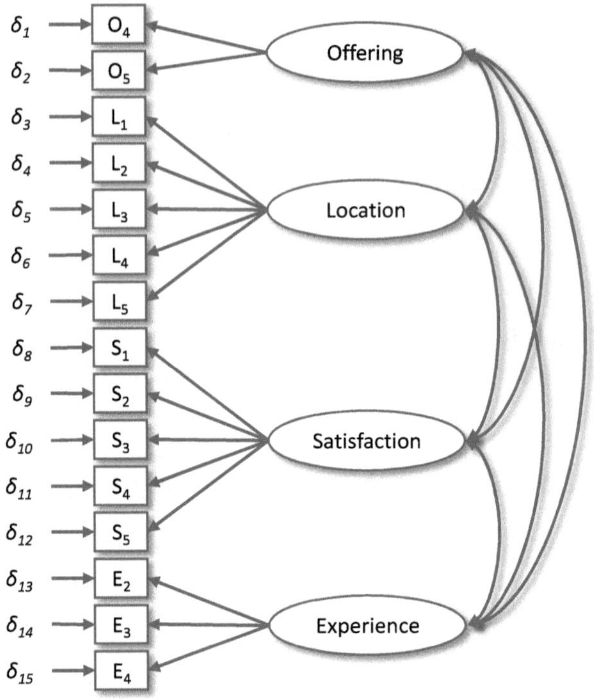

Fig. 12.12 Measurement model

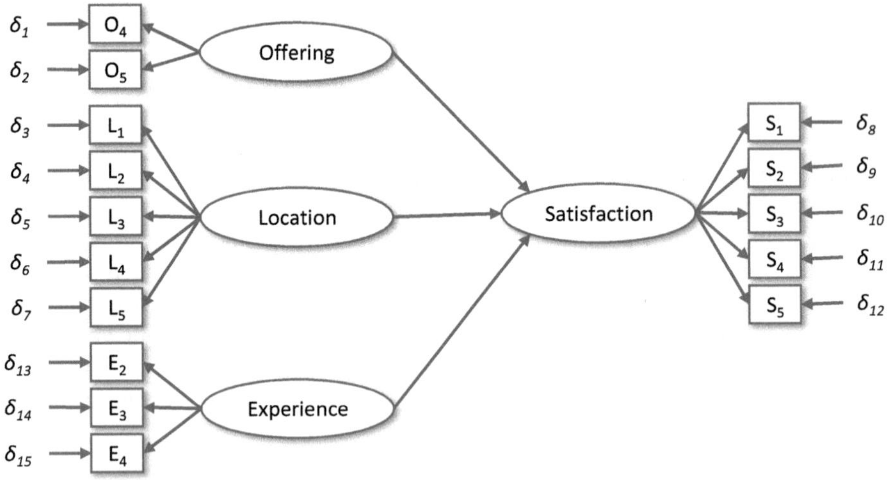

Fig. 12.13 Structural model

test based on the findings from the measurement model. The blue arrows between latent constructs represent the causal relationships. In the same way as regression, the structural factor loadings can be assessed with t-tests.

Confirmatory factor analysis is beyond the scope of this text so we stop our explanation here.

12.7 Summary

In this chapter, we described the data reduction technique, exploratory factor analysis. It can be used to reduce a set of variables (add them together) if they share a common factor structure. It is also commonly used to test construct validity of measurements of latent constructs. We focused on common factor analysis, but also briefly described two related procedures: principal component analysis and confirmatory factor analysis.

References

Anderson, J. C., & Gerbing, D. W. (1988). Structural equation modeling in practice: A review and recommended two-step approach. *Psychological Bulletin, 103*(3), 411–423.

Bollen, K., & Lennox, R. (1991). Conventional wisdom on measurement: A structural equation perspective. *Psychological Bulletin, 110*(2), 305–314.

Comrey, A. L., & Lee, H. B. (1992). *A first course in factor analysis*. L. Erlbaum Associates.

Jöreskog, K. G., Olsson, U. H., & Wallentin, F. Y. (2016). *Multivariate analysis with Lisrel*. Springer.

Nunnally, J. C. (1978). *Psychometric theory*. McGraw-Hill.

Spearman, C. (1904). General intelligence, objectively determined and measured. *The American Journal of Psychology, 15*(2), 201–292.

Tabachnick, B. G., & Fidell, L. S. (2013). *Using multivariate statistics*. Pearson Education.

Part IV

Reporting

Reporting Findings

13

Contents

13.1 Introduction

From a decision-making perspective, we have discussed both qualitative and quantitative aspects of the research process. Our mission with this book is to promote a data-driven scientific approach to decision-making. Our world is drowning in data. Data that is grist for the mill of data analysis. When properly analyzed, data provides a wealth of value to decision-makers. Given the nature of digital data, quantitative methods dominate the pages of this book. Despite this, we advocate for a pragmatic approach, fluctuating between inductively developing theory and deductively testing it. Qualitative and quantitative methods are innately related in a scientific process that leads to better theories and better decisions.

13.2 The Report

A report is written for the recipient!

Social scientists (and natural scientists) decry leaders who ignore their research. The literature abounds with speculation about why decision-makers ignore evidence. Caplan proposed the two-communities theory. Researchers and decision-makers live

J. E. Sallis et al., *Research Methods and Data Analysis for Business Decisions*, Classroom Companion: Business, https://doi.org/10.1007/978-3-030-84421-9_13

in "[S]eparate worlds with different and often conflicting values, different reward systems, and different languages (1979, p. 459)."

To communicate with managers, you need to understand managers. In Chap. 1, we briefly mentioned three basic models of decision-making: First, the *individual rational model* where managers collect and attempt to comprehend information that is pertinent to decisions. They are challenged by bounded rationality and psychological idiosyncrasies that limit and bias their decisions. Next, *the political model* where managers compete with other interest groups with conflicting agendas. Decisions are often conflictual with the stronger party dominating the outcome. Finally, the *garbage can model* where ambiguity abounds. Decision processes are unstructured and outcomes are often coincidental.

The responsibility is on researchers to bridge the gap between the research world and the managerial world. Approach it like a research question: *How can I communicate my results in a way that makes decision-makers listen and use the knowledge I provide?* If the recipient has a good understanding of scientific methods, then technical descriptions may be welcome. If the recipient has little understanding for scientific methods, then technical descriptions will likely confuse and alienate. Managers need to know *that* research findings are robust. However, they do not necessarily need to know *why* they are robust. Most people experience research methods and especially statistical data analysis as overwhelming. The magnitude of detailed information seems impossible to comprehend. We ask our PhD students, can you drive? Most say yes. Then we ask, can you explain the internal compression engine and drivetrain? Most cannot. Our point is, there are many things in life that provide tremendous value without having a deep understanding of their function. In the decision context, this means providing research-based scientific knowledge to managers in a way that they comprehend and trust. In the same way that they can drive a car without understanding its internal workings, they can make data-driven decisions without a deep understanding of the scientific method, so long as they comprehend and trust the science (the car).

13.2.1 Writing Style

Even a technical report can be eloquent in its telling. A research report is a story. Use straightforward and accurate language with the research question(s) at the center. Once concepts are defined, be loyal to the definitions by using consistent language. When explaining research methods, the most fundamental goal is to clearly explain the choices you made and why. From there, work in more technical expressions like validity and reliability. Do not talk about something you do not understand. You risk undermining your credibility. Examples and direct quotations from respondents help bring life to the narrative. Metaphors are useful for providing a perspective that the reader can relate to.

Figures and tables should assist the reader in understanding the content. They should summarize what is being expressed in a way that increases understanding. The adage, *a picture tells a thousand words*, can be used as a litmus test. Ask yourself, does the figure or table *replace a thousand words*? What purpose does it serve? Often, certain tables are expected as the convention for how to present specific information about statistical analyses. Look at how content is presented in the same context. If the report is aimed at or based on a specific academic journal, then look at that journal to develop your templates. Here is a list of bullet points to consider:

- Use sober objective language.
- Be consistent in how you refer to concepts.
- Use a consistent reference style.
- Avoid run-on sentences. Be concise and to the point.
- Do not use casual language (Good: do not; Bad: don't).
- Be gender neutral.
- Stick to conventions within the research context, like how findings are presented.
- Avoid overuse of italics, exclamation points, and other ways of highlighting text.
- Avoid humor; it is not appropriate in a research report.

13.3 The Structure of a Research Report

Research reports for public agencies often follow a specified template. The same may apply to private businesses. Always check first if there are specifications to follow. Below, we present a generic plan for how to structure a report. It is the typical structure for university-level research reports and theses. Writing the research report follows much the same order as the research project.

1. *Abstract or executive summary*: After the report is written, consider adding a short summery at the beginning: 150–300 words that summarize the research purpose, the research question(s), the theory, methods, and findings.
2. *Introduce the research topic*: With regard to writing the introduction, the topic should be clear and the reader captivated within the first few paragraphs. In a professional context, the topic is probably set. For a university report, the student can usually suggest topics. When you have a topic in mind, ask yourself two questions: Is the topic interesting, and where can I get data? We suggest looking at current topics in popular literature and online. For example, during the COVID-19 pandemic there is a lot of news about changing purchase behaviors and the impact on private businesses. Consumers are an accessible data source, and there are lots of research questions that are relevant and interesting for managers. Research questions that require getting data directly from managers are much more difficult. The reader needs to know what the research will contribute and possibly what theory is being applied.
3. *The purpose and research questions*: The introduction should end with, or be followed by, the *purpose of the research* and the *research question(s)*. In Chap. 2, we suggested considering what decisions will result from the research question. Faced with the new reality of life after COVID, managers will be faced

with decisions of how to adapt. Business-to-business relations have changed drastically under the travel restrictions. How will managers adapt to the new way of doing business?

4. *Limitations*: A research report should express the limitations or boundaries within which the research is undertaken. Sometimes, this can be presented as a section describing limitations. We suggest expressing them in a positive fashion as opposed to a negative fashion. For example, by clearly articulating the context, research questions, theory, and methods, the author is implicitly expressing the boundaries. Trying to express what the research is not covering is, arguably, infinite. Given this, express specifically what you will do so that the reader implicitly understands what you will not do.
5. *The context*: After the research question, there may be a section describing the *context* in greater depth than was possible in the introduction. Sometimes the context comes out naturally when developing the theory, so this section is optional.
6. *The theory*: Except in the rare extreme where the research is completely exploratory, the report should now have a section describing the *theoretical perspective* on the research question. In Chap. 1, we talked about ontology, epistemology, and research methods. We talked about the philosophical spectrum of constructivism to positivism. In many cases, like a managerial report, these concepts are implicit. In some academic reports, they are treated explicitly. Either way, they are relevant for how the research question will be approached, and possibly the theoretical perspective. At the exploratory (constructivist) end of the spectrum, the theory section may simply state what theory may be relevant for the research questions. In a more structured approach, the theory and data may together suggest propositions that the researcher presents. At the positivistic end of the spectrum, the theory should arrive at a set of hypotheses that will then be tested.
7. *The research methods*: In the research methods section, the researcher explains the methodological choices made and why. Its most fundamental purpose is to articulate the methods in such a clear way that other researchers could *replicate the research*, and thereby confirm the findings. It is the only section of the report that should be written in the past tense. Depending on the nature of the research question(s), explain the research design. When applicable, explain how secondary data was collected. Or, explain how primary data was collected. This can include things like describing the sampling frame or the criteria for choosing participants for qualitative data collection. Describe how things were measured. For example, how interview guides or questionnaires were developed. Explain how the data will be analyzed.

   Every time you make a methodological choice, there will be positive and negative aspects that weigh in. Some report formats include a section discussing critical issues, meaning the negative aspects, of the choices. For us, this is like dedicating a section to flaws. It undermines the veracity of the report. Instead, we suggest that you deal with the trade-offs each time they arise. That way, it is clear that you made the superior choice with the knowledge of it being a trade-off.

8. *Reporting results*: In general, the results section should be straightforward and with a minimum of interpretation. In other words, state the facts and what they mean. However, avoid making value judgments and speculating about deeper meanings. Reasoned speculation comes in the discussion or conclusion. If you are reporting results from, for example, interviews, consider including quotes to emphasize the findings. For quantitative analysis, there are many norms for how to present statistics, depending on the audience. If you are aiming for a publication, there will usually be explicit guidelines.
9. *The discussion and conclusion*: In the discussion, you have the freedom to discuss your results in terms of the context and theory. This is where you convey to the reader the deeper meaning in your results. Did the results turn out as expected? If not, why? Within reason, you can be speculative. Some authors write a relatively short discussion, putting emphasis on the conclusion section. This is a question of how you feel the results are best conveyed to the reader.
   As a rule of thumb, start the *conclusion* by restating the research question(s), and then answer them in order. If there are hypotheses, address them in order. It helps the reader if the structure established early in the thesis in the theory section is repeated in the conclusion. Though mundane, it reduces complexity for understanding the report. When applicable, include a section on possible venues for future research that emerges from the findings. Include a section on practical implications, or if the label fits, managerial implications.
10. *Appendices*: Sometimes in order to facilitate the flow of the report, things can be moved to the appendix. A common example would be to have the questionnaire used for collecting survey data in the appendix. However, avoid the overuse of appendices, and for that matter, footnotes. And, since we are being critical, be careful with acronyms. Always remember that you are telling a story. If something is important to the story, then it should be in the text. Imagine writing about Harry Potter (HP), and his friends Ron Weasley (RW) and Hermione Granger (HG). HP (footnote describing a 12-year-old boy) meets RW (footnote about the Weasley family), and HM (footnote about muggles) on the Hogwarts Express (see the appendix for information on Hogwarts).

13.3.1 Referencing

There are several reference styles available online. Often, schools and journals will specify which reference style to use. If you are using bibliography software, they often come with reference style templates. It is important to consistently use the style and to reference only those works that are used in the text. Especially for long reports, we highly recommend bibliography software like Endnote or RefWorks.

The reason to be so careful is for others, in a simple and predictable way, to be able to find the referenced material. This is linked to the verifiability of research. Every single comma, period, colon, and italic matters, so be accurate!

13.4 Implementation

Research is done to address a research question, and as is often the case, to support making a decision. Throughout the book we have presented many examples, especially related to the Hotel data. If the research shows that guests are more satisfied by having a selection of craft beer available in the hotel bar, then managers can decide if the cost is worth the benefit. If instead they find that there is no significant effect on satisfaction, then they know where they should not invest effort. Though this example is relatively mundane, if done well, this sort of research can greatly help decision-making.

To the researcher, the outcome and implementation may seem obvious. To the manager, it is more challenging. Imagine being the President of the USA. Just like any human being, they try to follow some sort of rational process for decision-making. At the same time, they are susceptible to the same biases as any person. Just imagine how difficult it must be to comprehend all the information they get (what we call, bounded rationality). The political forces around the decisions are immense. How can a researcher get the President's, or for that matter, any manager's attention?

Though we recommend avoiding humor in research reports, perhaps a bit of humor here will help make our point. One of Arnold Schwartzenegger's more poignant lines comes from *The Simpson's Movie* where he plays the President of the USA. When presented with five reports outlining his options, he immediately picks number three without even reading it. His rationale, "I was elected to lead, not to read."

While amusing, this is the reality for many decision-makers. Researchers compete for their attention. To have any hope of implementing research findings, they must be presented in a way that captures attention. They must be clear and credible.

13.5 Advice for Students

As a student, ask yourself, why am I doing this research?

We like to hope that all research creates knowledge. Even if the research results are negligible, hopefully the researcher has learned something useful. Undergraduate, graduate, and PhD student can take pride in contributing to knowledge through applying a scientific approach to answering research questions. At the very least, they have learned the analytical approach to problem solving. It is a valuable skill. We ask, why are you doing the research and writing the paper, report, or thesis?

To be blunt, you want to pass your courses and get your degree. How do you do that? By showing your examiners that you have the skills that qualify you to deserve the degree.

Be focused. Clarify what is expected from the project. If it is a focused topic with specific parameters, stick to them. Occasionally however, you are given a broad palette upon which to show what you can really do. This is often the case with the final paper or thesis of a degree program. In this case, consider what is the best possible way for you to demonstrate your skill set. We suggest using multiple methods. Stretch yourself, but do not overextend yourself. You have limited time and resources, and you have a limited skill set. If you go too far and overextend time, resources, and/or skills, you will likely fail.

Having said this, if you have confidence in your skills, dare to try multiple methods. Choose a research topic where you can start with a qualitative exploratory research design, and then test the findings with quantitative methods. If you do questionnaire survey, then you should do an exploratory factor analysis and reliability analysis to validate the measures in the questionnaire. This would be followed with a method to test your hypotheses. If you are comparing groups, use some sort of *t*-test or analysis of variance. If you are hypothesizing dependency through causal relationships, then use regression analysis. Do not bury the story in analyses. Instead, use them to tell the story.

13.6 Summary

Reporting research for decision-making is to overcome the communication gap between two communities: researchers and decision-makers. The responsibility is on the researcher to overcome the gap and get the attention of the decision-makers. Pay attention to conventions for how to present research in the context you are reporting in. The report is written for the recipient. If it is for a technical audience, then use technical language. If it is for a managerial audience, then use managerial language. This chapter outlined a generic set of steps to follow when reporting. The emphasis on each step depends on the research and the audience. Remember that in a world overwhelmed by information, communicating your research is a competition. The value in research is created when it is used, not when it is conducted. Be clear, concise, and ambitious.

Reference

Caplan, N. (1979). The two-communities theory and knowledge utilization. *American Behavioral Scientist, 22*(3), 459–470.

Index

J. E. Sallis et al., *Research Methods and Data Analysis for Business Decisions*, Classroom Companion: Business, https://doi.org/10.1007/978-3-030-84421-9

MIX
Papier aus verantwortungsvollen Quellen
Paper from responsible sources
FSC® C105338

If you have any concerns about our products,
you can contact us on
ProductSafety@springernature.com

In case Publisher is established outside the EU,
the EU authorized representative is:
Springer Nature Customer Service Center GmbH
Europaplatz 3, 69115 Heidelberg, Germany

Printed by Libri Plureos GmbH
in Hamburg, Germany